Automated

The concept of continuous results

with diminishing involvement

Joba Adekanmi

Automated

ISBN 978-1-914528-01-9

Published in United Kingdom by:
Impact Publishing House

www.TheServantandKing.com
iph@TheServantandKing.com

Contents

Preface   5

Introduction   11

Chapter 1:   Simple or Complex?   21

Chapter 2:   The Search   33

Chapter 3:   Behind the Scenes   47

Chapter 4:   Learn from History   61

Chapter 5:   Opportunity   79

Chapter 6:   One-Touch   91

Chapter 7:   The Value   101

Chapter 8:   Think Automation   115

Chapter 9:   Creativity   127

Chapter 10: Diminishing Involvement   141

Chapter 11: Some Concerns   151

Chapter 12: Some Solutions   155

Chapter 13: A Product of Time   163

Chapter 14: The Future or the past   173

Chapter 15: The Right Choice   189

Contents

References   193

Preface

We have heard the famous quote: 'Doing the same thing over and over again and expecting different results is insanity.'[1] It may imply that doing the same thing over and over again and expecting the same result is sanity. I wonder if it is acceptable to be true.

I found a transformational truth that affected me, something more unsettling that gave me insight. It is a fact I discovered in the course of my work. 'What is it?'; You may ask. It is doing the same thing repeatedly and expecting the same result could be the greatest destroyer of potential.

You have done it once, you did the same thing again, you expected the same result, and you got it. It looks good. Right? If it is the same each time, you are not making the best use of yourself and your potentials.

There is a principle that can help you. I have figured it out; it is the power that is changing our world at such a rapid pace. It is the energy that enables you to reach more with less. It is the strength that helps you do more. It is the power of automation.

I wish many people understood and applied it to their lives. I desire that many more will get to know and embrace it. It is a game-changer. Doing the same thing continually yet expecting the same result is the greatest destroyer of potential. If you do not understand this principle, you may be achieving less than your potential.

The technology giants and many big corporations have flourished in a short time not because of luck but because of a principle that allows you to do more with less. It is a stunning concept that works and has been with us for a very long time.

Google, for instance, is providing a service that is used all over the world. The average number of queries processed every second is over 63,000. It translates to over 5.4 billion searches per day on average, and it is still increasing.[2] How many employees would you have thought would be able to sustain such a global service? The number of employees is less than 0.002% of that daily search.[3]

YouTube, which operates as a subsidiary of Google, makes available over one billion hours of content for people to watch every day. The online video sharing platform enables billions of people to watch video content at the same time from different locations around the world.[4] There is a principle behind that ability to reach many with few. It is a concept they have found and made the most of it.

There are so many things that happen, but you cannot explain why and how they happen. We see so many results, but many people never pause to understand the input that produced such startling outputs.

What is the reason behind the action? What is the cause behind the effect we see? The answer could be what you have been waiting to discover. Do you want to do more with your time? Do you want the same results even after your last action? You may be closer to the key that will unlock the truth.

Automation is simply a principle of creativity, allowing continuous results with reduced input and activity from you.

We all live on this planet called earth as human beings with over 7.8 billion people. We have all enjoyed the benefits of automation. We still do. I found this to be true for all people across all nations; that has ever lived or will ever live. Many people, however, do not understand this truth. They live but are still yet to make the most of their potentials. It is a great tragedy that only a few people understand and apply this principle. It would be helpful to produce results worth more than twenty-four hours of activity within twenty hours.

Have you discovered the value of time and now wish you could have more than twenty-four hours in a day? Are you still struggling to fit all your activities into the twenty-four-hour box? Do you believe you have potentials, convinced that you could do more, but the time is not just there?

This book may be the help you need, as it will expose truths and principles that could transform your life. You will discover that it is possible to make and allow things to work without you being there, yet achieving the same result or even better than you would if you were there to keep doing it.

You will also find out that you can recover your time and use it wisely while you keep getting continuous results even after your last action.

I hope this will be of help to you. I invite you to discover the power of automation, how the understanding and application of the principles can help recover your time and life regardless of your background.

Welcome.

◆◆◆◆❖◆◆◆◆

Doing the same thing
continually yet expecting
the same result is the
greatest destroyer of
potential.

◆◆◆◆❖◆◆◆◆

Introduction

Have you ever thought of the things you have to do daily and how many of them you have to do, again and again, the same way?

Everyone does something, and many people will have to do the same thing again sometime afterwards. It could be minutes after, hours after, days after, weeks after or even years after the previous action. It could be any activity relating to you, your work, the services you provide to other people or what you produce. Consider it carefully; is it the same input and activities every time, with the same and corresponding output on every occasion? Do you have to go through the same process every time you want to produce that same result? I wonder what your answers would be.

Imagine you invested so much effort and time to complete a task or series of actions, and you were so pleased by the results you achieved. You had a feeling of completion and a sense of accomplishment. It does not usually stop there. You now want to make it happen again. What would you do? Start all over, doing what you have once done. Yes, again. How does that sound?

I believe you must have done something that brought you your desired expectation. You completed it with joy at the first instance, but now you have to go through the same process to keep getting that same result you have once obtained, and you expect to get the same every other time you perform the actions. I guess you have some words to describe a scenario like this one.

It could be a process that produced information, manufactured goods, provided services or probably one that provided input for another operation. Yes, anything that requires you to redo your actions, process, efforts, observation, confirmation, testing, and so on. It is a continuous process.

You may well be achieving the expected results all the time in line with your desires. However, the effort you are putting in could be draining you because you need to do the same thing repeatedly to get the same result. It may not mean much if it were a simple process or the time involved is negligible. It is disturbing to know that many people wear themselves out to achieve the same thing. I know that so many people dread the feeling of having to do again what they have done once because it is complex, demanding, exhausting, tiring, complicated, prone to error, or frustrating. They have continued in some cases, achieving less with more. What is the way out?

Impressive goals achieved will require more to keep the same outcome and produce more. Consider a product that took so much time to complete, but you were satisfied by the result. You now desperately desire to do more but time, as some say, is not their friend. I can do more, you say, but there is no time. It is a desire that a day has 30 hours instead of 24 hours, a wish for the clock to come to a pause or a hope that the speed at which it ticks drastically reduces so that you can get more done in the same space of time. Have you ever found yourself in such a situation? You know what to do, you know how to do it, you know when to do it, you know what the outcome of all your efforts and actions would be, you are happy with the result expected, you are glad to put in all the effort required, but time is just not sufficient. What would you do?

At some point in your life, you must have written a test or examination or some form of assessment where you had a fixed time to complete the required tasks. You may be preparing for one at the moment or have some to sit for in the future. There are so many of these assessments that we face from childhood to adulthood. It could be school, trade, work, safety, professional, compliance or any other in life that evaluation is required. Everyone participating in an assessment will usually have the same amount of time to complete the tasks, and the time allocated is finite.

There was a Council report that revealed some findings on student testing. According to the report, a student takes an average of 112 tests between preschool and high school in the urban school districts surveyed. It also found that students sat for Tests over 6,570 times in one school year.[1] It will be practically impossible to go through life without understanding and facing the challenge of the limitation of time.

Time is limited for any test or assessment. It may not be pleasant to be in an examination hall when you do not know what to do. Boring it would be if you have no idea of what is required and frustrating if you do not know the answers.

Can you imagine a situation where you know all the answers, but you cannot complete the tasks in the limited time set? That is not good news. It could even be more painful, frustrating and discouraging. If it translates to failure or poor performance, there may not be a difference between the former and latter student. The outcome or success, in this case, is the ability to deliver on time; otherwise, the result will represent a lack of the required capability to succeed.

Life could be that way. You may know what to do, how, and when to do it. That sounds great. You may rightly predict the results that would attend to all your efforts and actions. It still depends on one major factor, the ability to complete all required before that last second of the test. Unfortunately, it is not always the reality for so many people today as they still have more to do when the time is over. Whenever the task required does not complete on time, the result will be different because the available or set time would have elapsed. How can you maximise time, fit more in, or do more in fixed time?

Have you ever completed the same tasks or series of activities and desired that the results were the same? You can discover some tips that can help you with some of the things you have to do and engage tools that will work for you to produce the same or better results with reduced time. This concept can increase your productivity and allow you to reach more people. It can empower you to do more with less, producing the same results even after your last action. You may finally be able to recover your time to do more crucial and often neglected things in your life that you may have long desired to do. The good news is that it will not be at the expense of your results.

Have you wondered why some organisations produce phenomenal results with fewer people and in less time? 80% of the world uses YouTube today. Two billion users can access this video platform from anywhere, through the internet, to view over 400 hours of new videos available every minute, plus over 800 million hours of videos already on it.[2] This service started fifteen years ago; today, four out of five people in the world make use of it. They were able to overtake many organisations within a short time, reaching more with less. Many more have figured this out and are doing more with less.

You may finally discover the power that is changing the world we live in and how many people, organisations, and nations are taking advantage of this concept. Many are still wondering and have no idea or clue as to what is happening. Many people have lost their jobs and means of livelihood; they are still wondering why. Some people seek relevance and employment with some expertise or skills; they seem to be frustrated as they cannot make headway because their skills are no longer required. There is a constant complaint of more to do, but far fewer skills, time or money to get it done.

Nations and organisations are going backwards instead of forward economically. They cannot figure out the reason. Many national services cannot cope with demand, as the skills required are in short supply or sometimes not even available within reasonable reach. Big businesses are folding up as they cannot compete with the market. New services are gaining ground and replacing many that have thrived for generations. Some still wonder and sigh; what is going on? What is happening? Relevance and prominence have shifted from the traditional and well known to the untraditional, new and uncommon, but unfortunately, many people are still not aware of what is happening.

I will like to introduce you, in the simplest way possible, to the world of automation. Regardless of your background, I hope this will be simple and easy enough for you and anyone to understand the power of automation. I desire that you will be able to apply the concepts to make the world a better place, thus making the most of life.

I hope you will find this insightful and helpful.

CHAPTER 1
Simple or Complex?

Interestingly, for the few or many people that have come across the word automation, it means different things. Many people are not sure, and probably a lot more do not have a clue. What is it, and how can it be described? Is it robots, or computers or information technology or artificial intelligence or one of the brilliant stuff that does strange things, as some may put it? Is it just something for some few in a professional discipline or only useful in a particular area of work?

Some people have asked, and many more are asking, much more would still desire to know what it is all about. Let us attempt to unravel the mystery behind the word.

Most often than not, the length and complexity of processes make it desirable to find something that could replace or reduce the human effort involved. The process could be initiated or completed by an approach or method of working that uses electronic devices to achieve the same result. It could be a device with a mechanical makeup that also has some electronics to provide some intelligence to manage and control it. The configuration is such that all possible decision making required throughout the process is incorporated. Eventually, it replaces the human controller or operative that interacts with it from time to time.[1]

In the production of goods in large quantities, some appliances used can control all or part of the process such that less interaction is required. These activities were done by humans partially or fully before the advent of the relevant technology. The suitable alternative in part or in full has made the work less laborious for humans and increased productivity. The solution may differ in various areas of life, but it is all still achieving the aim of making the work easier and faster.[2]

There are some devices or computers built to perform complex functions. They could do the work done by human beings during the process with little or no interaction. Many do not require control as they could operate and complete the tasks they have been designed to do, from start to finish.[3]

There are so many machines that have helped reduce the energy required to do work. They have helped increase human efficiency and effectiveness. The ability to control these machines has also been made possible. They could benefit from a unit made of various devices that forms a system to manage, coordinate, and command its activities and the related components.[4]

There are a lot of descriptions that can help one understand more about automation. Everyone can provide a solution that could simplify their work in their area of life. Some people avoid it or anything to do with it as it usually seems related to only technology professionals. Some people will not give any attention whatsoever because they think it is not for them, or probably it has been seen as an enemy that has once affected their life negatively. It is perhaps the reason for the loss of a job or business.

A simple explanation of the well written and technical definitions you may have read is not far-fetched. Simply put, it is; let something else do what you do continually. In other words, you do not have to do what you do all the time when you can get something else to do it for you. You do not have to do the same thing continuously. Let something else do it for you while you engage your time in what only you can do. You can easily understand this concept and apply its principles to everyday life. As much as possible, let something or someone else take your place to do what you do continuously so that the work can be done without you, and then you can do what only you can do.

We all have something we do regularly, and we spend time, energy, and resources to do it. If we can get something else to do the task, then we can automate. Automation is actually for everybody, and apposite for any area of life. The doctors, farmers, nurses, carpenters, pilots, engineers, any living human being can apply the principles to whatsoever they do. If only you can understand the concept, you will discover that it is for and can be enjoyed by all. I hope I will make it simple enough for anyone to understand and make the most of it.

I hope that you can grasp this as a life approach to solving problems by engaging the concepts to continually achieve the same or better results than you would have achieved by your full continuous involvement but now with diminished input. Free yourself as much as you can. That is the idea. If something else can do it, why waste your life when you can do more with your life. That is why everyone can benefit from this concept if only they understand and apply it to their unique situation.

It is common to see its uses in solving highly complex problems which require high-level technical expertise. It is apposite for any life situation or field of human endeavour as long as there is the willingness to do what is takes or seek help and information that could make the journey to achieve the specific goal a success. It is vital, however, that you have a good understanding of the advantages and your aim. It could make your life more efficient as you carefully consider the viability on a case by case basis and make decisions based on your priorities.

Have you ever wondered why some people have a lot to do and still get everything done at the right time? Many have been amazed to see people that manage their activities well. They know what to do and the time to do it without tasking themselves to remember. 'How are they doing it?' A question that many have asked and some are still struggling to figure out.

Let us takes some simple day to day examples of activities you would probably like to do with some preferences.

Imagine you would like to remember to do the following;

Call the taxi, but only ten minutes before you leave home.

Call friends and family, but just 6 hours after daybreak on their birthday.

Leave for the airport, but precisely three hours before your flight.

Start or complete your assignment, but only seven days before submission.

Book your MOT test for your car, but exactly two weeks before expiry.

Leave home for the gym, but just 10 minutes before your time slot.

Pay your bills, but only five days before it is due.

Visit a friend, but precisely three hours after he is available or back from work.

Visit the stores, but only 4 hours before it closes.

Pay your house rent or mortgage, but on the day it is due.

Cancel your contract, but precisely one month before it ends to avoid charges.

Move your car, but precisely 5 minutes before the ticket expires.

Have you figured out how some people may be able to do some of the actions listed above? You have always thought they are smart. Yes, they are. They are smart, that take advantage of relevant tools or devices to help achieve their goals. It also reduces their involvement. You could also take advantage of some tools or make some so you can be smarter than you are or make others smart too.

These are familiar tasks carried out in different places around the world every day. They can benefit from a simple solution that would make it easier and less dependent on the person doing it. It is true. You may have found out too.

Simple devices that most people possess today can help save time and free them from unnecessary burden. An alarm set on a clock, phone, tablets, laptops, desktops with relevant applications can serve as reminders for any time of the day. It works.

A calendar application on different platforms can record events or activities. They have reminders that work for any day of the year, even for many years in advance. It will do the work for you while you keep your mind at rest, and you would not miss anything important.

You could apply these principles to do simple everyday actions by using alerts and reminders. It could be through a device, tool or even sticky notes. For instance, you can place a sticky note in front of your fridge or dining area. It can serve as a reminder for your activities or events. It could do something automatically for you every time you visit the kitchen. It would remind you of the event or activity you have written on it.

Every time you visit a location, you could be reminded of an event or chore or task because you have placed a one-time obvious sign there to do the work for you continually. You are taking advantage of something outside of you to help you remember the task every time as you have pre-planned it once. You do not need to do the work of remembering. It has been done for you outside you and also continually. Yes, you do not have to carry too much on your mind all the time. It helps you to be relaxed and focus on other things without the fear of forgetting and missing the tasks.

Do you have to remind children of their tasks all the time? You are not alone. I can understand your frustration if you have to do it over and over again. It could be draining and tiring. I found a simple solution. I discovered that there are some things they never forget to do. One example is that they never forget when it is time for food. They remember to visit the kitchen, so I do not need to remind them. They always remember to get what they need from the fridge when hungry. I, therefore, decided to print out the information they often forget on paper and placed it in the location they remember to visit. It could be on the door and or fridge, and they get to see the reminder every time they visit that location. I did not have to do it anymore; the printed paper is doing it for me. It is simple but very effective. You can also find out what is wearing you out and find a solution that could help you maximise your time.

◆◆◆◆❖◆◆◆◆

Free yourself

as much as you can.

◆◆◆◆❖◆◆◆◆

CHAPTER 2
The Search

There are many places around the world known for certain things. The United Kingdom is famous for so many things, one of which is tea. In the 18th century, it was a luxury and perceived as an upper-class product. It became popular and attracted more demand from the people. In the 19th century, the people witnessed the emergence of the electric kettle. It helped to satisfy the increasing demand of the people. Many homes, therefore, had a kettle that is used for boiling water to make tea.

To make tea, you had to boil water in the kettle, and you had to keep checking until you think the water is at the right temperature. You can then switch off the kettle and use the boiled water to make tea. If you did not switch off the kettle in time, the water starts to evaporate and reduce. Should you leave it for longer, you lose more water. If you forget about it or go outside for the day, then it could become dangerous.

The simple process of tea making could end up being a disaster that could burn the entire building. I guess you would not want that to happen, do you? 'No, thank you.' I know, but what do you do to prevent it?

You may decide to keep checking the water, time and time again, till you are satisfied it is at the right temperature. You then switch the kettle off and make your tea.

You may have observed that all the actions involved in the process will usually remain the same and that you will keep doing the same thing over and over again each time you want to make a cup of tea. You will keep checking the water, moving from upstairs to downstairs, from the bathroom to kitchen, from study to kitchen, standing in the kitchen waiting instead of doing something else, interrupting your work to check and so on. You will repeatedly spend time and energy to check that the water is boiled and then decide to switch off the kettle.

Can you imagine a situation where you do not have to do any of those repeated actions and still get the same result all the time? If you think of it carefully, the following will always be the same:

The boiling point of water is 100 degrees Celsius or 212 degrees Fahrenheit.

The water will be heated and will increase in temperature.

A tool like a thermometer can check the temperature of the water. Observing proven signs can also be used to determine when the water is at the desired temperature.

The decision you want to make; to switch off the kettle when the water temperature gets to boiling point.

If these factors will always be constant, then what is the value in trying to use your time, energy and resources to keep making the same decisions that you have already made once? The time, energy and resources could help in doing something else or conserved to focus and be more productive in making new decisions. It is a win-win situation. Even though you would be less involved, that does not significantly affect the result you could have had when you were fully engaged because you will still have your tea.

There are so many ways to reduce your involvement in the process. One is to get some information from the process and see if you can suitably replace your repetitive involvement. You will need some information to make decisions and some tools to help you achieve that goal. Let us consider some possible scenarios.

Scenario One:

You can experiment to determine how long it takes to boil water and use that time as a guide every other time. A thermometer can measure and indicate the temperature. It makes it easy to confirm when the water gets to the boiling point of 100 degrees Celsius or 212 degrees Fahrenheit. You can make a note of the time taken for future use. For instance, if you found from your experiment that water boils in three minutes, you can use it as a guide every time you repeat the same process using the same parameters. You can then engage a device like an alarm to alert you after three minutes. The result of the one-time experiment can save you from the continuous activity of trying to determine the right temperature of the water by guessing, feeling, or checking continually. So you have replaced your involvement with an alarm that alerts you at the right time. It is a confirmation that the water has reached the desired temperature, which prompts you to switch off the kettle and make your tea.

Scenario Two:

Let us think of some of the tasks and the possibility of using a different type of device(s) to reduce human effort. The next step is to search for what can do what needs doing. It will be helpful to have something that can measure and determine when the water reaches boiling point. Another that can disconnect the power or switch off the appliance at the right temperature will be great. The next question will be; Are there any familiar device(s), or can new ones be made? A thermostat can sense the temperature changes and initiate the turn off at the desired temperature.

There is a device called a bimetallic strip used in thermostats. It is an electrical contact that is temperature-sensitive and can convert a change in temperature to mechanical displacement. A combination of these devices can perform the work of a thermometer and control. If all the suggested replacements can function as required, it implies that it is possible to use something else capable of doing the same thing all the time. It can all be set-up to form an appliance that would require less attention.

Once it is set-up, the required action would be to switch on the appliance filled with water. It already has everything built-in to carry out all the other tasks that you would have done, and you come back to meet boiled water and an appliance safely turned-off for you, and then you can make your tea. That sounds less involving, time-saving and efficient.

Your continuous involvement in repeating the same tasks and decisions is no longer necessary. The different components that can do the job repeatedly handled it for you. It implies that you can use your valuable time, energy and resources to do some other tasks. That is the idea of automation. The appliance I described is what we now know today as the automatic electric kettle.

In the 20th century, we saw the advent of the innovative K1 kettle, which safely powers off once the water reaches the boiling point. This invention was the world's first automatic kettle.[1] Someone did this work, and the world benefits from it today. It has saved many houses from burning, reducing disasters, saving time, energy and resources. Simple solution with numerous benefits, that is automation. Everyone can; yes, you too can. We can all do something to make the world a better place.

A British entrepreneur, John Taylor, produced a bimetallic thermostat that is smaller than a fifty pence coin. It could easily fit and function in the kettle and used by kettle manufacturers. He provided a solution that could replace the continuous involvement of the users. It has become a global solution that is used over a billion times every single day.[2]

The first scenario also implemented automation, but the second had less involvement. You can even go further by automating the other processes of making the tea and delivery and so on. The former has more risks as it depends on you to act. You may forget or decide to ignore the alarm, but the latter completed the required tasks safely without further involvement. There is no end to it. It is as much as you want or as far as you decide to go. If you keep doing the same things and you expect to achieve the same result all the time, it may be worth considering using something else to do the work.

It is all about reducing your involvement in the process. You could start from what you do daily, regardless of your profession, sex, race, nationality or education. If it is a process, action or activity, and you repeat it continuously, the concept may work for you. Do not be fooled; you can make it work.

Many people have the task of providing the same information to people continually as part of their duties. It could be in a fire safety class, new starters or leavers in the workplace, visitors to a site, commuters in a bus, train, aeroplane or ship, training course or children's frequently repetitive activities. If it is the same thing all the time, you can benefit from this time-saving concept.

The information could be produced once by the desired trainer and recorded with an appropriate tool in the desired format, like audio or video. Whenever the same information is required, it can then be retrieved on-demand. The same information is then provided continually without the involvement of the initial trainer. Remember, it is also achieving the same result of educating the listeners. The Speaker or the expert does not have to be there anymore. If required, anyone, even without the slightest idea of the subject, can push a button to reproduce the same result, thus reserving the high-level skill for a more important assignment. If there is no such need for personnel, those that need the information can serve themselves. It can reproduce the same result all the time, regardless of the person that activates the device.

I know we are all different, and we have varying interest and preferences, but whatever you do, you can apply this principle to improve efficiency and effectiveness. We all make decisions, and some of them will always be the same. We take actions in certain situations that are most often than not repeated in the same way. Anytime you open the door to a dark room, for instance, you would like to illuminate the room. It is likely to be your choice every time so you can see the things in the room and not stumble on an unknown object.

You will have to do the following many times: Locate the switch if you already know the location or try to look for it, if you are not familiar with the room, then you may try to apply some common sense, building knowledge, former experience and other information sources to locate the switch so that you can switch it on and then remember, yes remember to switch it off when you are leaving. You know the situation, and you know the decision you want to make in that situation. You can get a device to do that job for you all the time. You only need to open the door, then the device senses and detects that someone is there and then switches on the light for you. When it cannot sense any movement, it switches off the light for you. It continues to do this repetitively every time it is triggered.

You will be shocked to know how many people switch on their television and forget to turn it off or sleep off while watching a programme. It is watching them incur bills instead of them watching the television. They have to pick up the bills for what did not benefit them. Many leave water running and forget to switch it off on time.

For some others, it could be the computer, cooker, boiler, air-conditioning, fan, oven, car headlights, security lights, and just anything that you can think of that needs action from the user. They have forgotten to initiate it or probably did not do it on time. The cost could be as little as a few damages costing some time or money, to dangerous damages that could cost more and probably mean being out of pocket because of a high bill or emergency expenses.

Examine what you do daily, and you would find out that you are likely to do many of them more than once. Think of the following actions and consider what could help with the tasks;

Parents instruct their child every day.

You read the same or a few books to the same students every day or different students every year.

You read a storybook to growing children.

You conduct orientation for a new starter or leaver at your workplace every day.

Teach children how to brush their teeth.

Train new employees on how to use some software.

Conduct employee health and safety sessions.

Organise site induction for new employees.

Show visitors around a site every day.

Read safety instruction on the plane, bus or ferry.

A device that can do any or some of these actions for you can help you. You are reducing your involvement by doing so. You are taking advantage of this concept by automating your processes. When you discover what can help you and engage it correctly, you will be free, and the work will do itself without your involvement.

CHAPTER 3
Behind the Scenes

It was one of those holiday adventures outside the country many years ago. A high school boy and his family were visiting a favourite country overseas for the first time. The journey was eventful; there was so much to process within a short period. Wishes have come true and now faced with a unique opportunity to explore a different city and find out more about the people. They stayed in one of the cities, with a planned tour as much as the time they had would allow.

It was time to go out for some shopping and sightseeing; they were all eager to go, so they followed the directions they were given and made their way to the shopping mall. The young man leading the way soon found a popular store. He decided to make his way to the entrance, moving at a fast pace, one step at a time, while others followed slowly, observing and enjoying the beautiful scenery.

As he approached the door, he wondered and tried to figure out how to open the door. He looked carefully, observing the sides of the wall to locate the door handle. He continued yet slowly, wondering how he would gain access. Suddenly, he heard a sound and saw the movement of the doors as they opened. It seemed something out of the ordinary has just happened. 'The doors opened without any human action. Strange!' He said to himself. He ran back quickly to his family to break the news of his discovery.

Amazed, surprised and searching for answers, he desired to find out what the mystery was. The planned holiday soon turned out to be another opportunity for discovery as he searches for answers and seeks to find out the missing part of the puzzle. He wonders, 'How on earth can this happen without doing anything to make it happen? What is happening here? Is this magic? Is there a ghost monitoring me or trying to scare me?'

He saw one security personnel approach that listened to his questions and smiled in response, 'None of that, I am afraid. It is automatic.' They later walked into the store through those doors for their shopping.

Yes, those doors are known as automatic doors, it opens and closes, but you never really see anyone performing those actions. It works in a certain way and responds to triggers. It is not magic but expected. Those that understand how it works will appreciate that it is doing well.

They continued with their plan afterwards, and he began to notice more of the same type of doors in different places. He also found that he had a similar experience every time. 'The doors are in so many places.' He said. He found them on trains, stores, airport, workplace, surgery, post office, and so on. He was not satisfied with the simple answer he got. He wanted to know more as he wondered, 'What is the secret to this mystery?'

'How does it work?' He felt confused. 'Not all doors are like this, and not all behave this way. Probably, something strange must be going on. Doors should not open except you open them. Who told the door to open? Why should the door open? How did the door know it was time to open anyway?' He asked these questions with interest in the workings of the doors as he sought answers.

He had seen so many doors that were not automatic. An effort is usually required to open or close it. He could recollect the actions, and energy dissipated. He would usually approach the door, grab the handle to open, then pull, push or slide, depending on the type. That is how he has known it to work. There is nothing strange about that process because he is used to it. Oddly, that process did not happen on this occasion, yet the door opened. That is why it is a bit suspicious.

He thought of a good use case for the automatic doors in busy places. He considered the problems if the doors were as he knew it. It probably may not mean much if you only had to open and close the door of your house a few times a day. You may have to perform that action when you go out and when you return.

Let us take the case of a grocery store as an example. If someone had to open the door for every customer that visits and leaves the store, that would be so much effort. If it is hefty, having to push, pull or slide it many times a day will not be fun. Can you imagine having to push, pull, or slide a heavy door open 1000 times or 2000 times in a day? You may not find it exciting after a while. It would be helpful if you could get it done without having to dissipate so much energy. The time consumed all day can help to do more productive work.

Today, we may take the modern automatic door for granted. It was not until 1931 that The Stanley Works Company installed the world's first automatic doors. It was in a restaurant in Connecticut.[1] It is only in the last nine decades that we have come to expect doors to slide open for us in offices, shops, airport, gyms, and other public places. Some stores have over three million customers through their doors every day. They have implemented a solution that could give their customers a great experience. They can easily walk in and out of the store because the doors open and shut automatically. It saves customers' time and energy.

A Lesson in Science and Technology

Many may still wonder, just like the young boy. 'What is it that has made this possible?' A closer look will always give a clue. If you move closer and observe, you are likely to notice some things when the operation starts. You will probably see some of the moving, non-moving parts and perhaps hear some sounds. Some things contribute to the working of this process that is out of sight. If you do not know and cannot see them, you can always find out instead of assuming.

Scientists have worked so hard to explain what some people could still refer to as magic or mystery. There are indeed so many things hidden that make it work. Many actions performed that are not visible, as they happen in the background. Its various components are usually covered, some under the ground or between walls, although some could be visible but may not obvious.

Sensors are part of the operation of automatic doors. As the name implies, they can detect many things like sound, light, weight and motion. Different types exist to achieve the same result. It made it possible to know that someone was approaching the door of the grocery store.

The automatic doors operate when a sensor sends a signal. They perform the required action of opening the doors upon receipt of the sign. The sensors could be under a rubber foot mat in front of the store to allow customers to step on it. It relies on the assumption that everyone will walk past it as they walk towards the store. You only see a mat that you would probably assume is just for you to wipe your feet, and never think it has another function. You step on the mat as you approach the doors, don't you? Yes, It is normal. The sensor can sense your weight when you do so. It sends a signal that activates the operation of the other mechanisms. The door opens as a result of the cumulative actions.

Behind the Simplicity

Let me attempt to make this a little simpler and not too technical and complicated. Let us list all the actions that will be involved in entering any store, for instance. Let us also assume that one security personnel will open the door for you, so you do not have to open the door to simulate a similar experience.

You will do the following;
Approach the Store.

The security personnel will do the following;
Sight you as you approach the door.
Locate the door handle.
Pull the door open to about 90 degrees.
Or
Push the door open to about 90 degrees.
Or
Slide the door open to about 1 meter away from the wall.
Ensure the door is kept open with enough room for entrance.

You will do the following;
Walk into the store.

The security personnel will do the following;

Wait till you walk past the door.

Close the door behind.

You can see all the steps and actions necessary to be carried out for you. All you need to do is just to walk-in. Let us attempt to automate these tasks so that all you need to do is to approach the store and walk-in.

I will need to replace each step with a suitable component capable of acting the same way. A single device that can do all will be great, but it is rarely possible. Ideally, two or more having different functions will be required to complete the tasks partially or fully. The order of activity can then be determined and set to achieve the desired goal.

Relate the actions to the possible replacement components required.

Action: Sight you as you approach the door.

Component: A   sensor   can   detect when you walk towards the door. That is one component.

Action: Locate the door handle.

Component: A mechanism fixed to the door that would help push, pull or slide.

Action: Pull, push or slide the door.

Component: Belts, motors, track, wheels could make it easier to move the door

Action: Power to Push or Pull or slide.

Component: Power Supply Unit and Battery.

Action: Ensure the door is kept open with enough room for entrance.

Component: Mechanism to open to the desired length.

Action: Wait till you walk past the door.

Component: Sensors   to   detect whether there is still someone there or a timer to allow sufficient time to walk past.

Action: Close the door.

Component: Control   Unit   with electronics. The sensor activates the closure when it cannot detect or sense anyone or activity. It can also close after a set time.

There seems to be a possible replacement component that could perform most of the actions. They are sensors, belts, batteries, retrofit kits, automation accessories, wheels, control unit or controller, the power supply and battery.

All the components that perform one or more of the tasks required are now brought together and assembled in the best way possible to achieve the desired result. The result will be that the components perform the actions, and they could do it repeatedly.

Automation has taken care of all the activities. A walk on the mat is all that is required to activate the process. A sensor will detect that signal, and the process begins. It will continue to run that way without any other human effort required.

There are so many use cases for automatic doors, and that is why we find them in numerous locations and different types. They are in supermarkets, post offices, large buildings, trains, among others. They make it easier for people to enter and exit through the door.

Automatic doors are of great benefit to people that have some disabilities, especially those that prevent them from doing some or all the actions involved in the process of opening the door. The weak and the elderly enjoy the benefit as well. Those taking care of children using buggies would find it helpful. If you ever find your hands full when shopping, you would see an advantage in this solution.

It is also energy-saving as it conserves energy, heating or cooling depending on the environment or season. The doors only open when triggered by someone that needs to enter or exit the building. It helps to conserve energy as against opening the door all day. It is a solution that saves time, effort, resources and solves other problems too.

Automation reduces the effort and the number of tasks that one needs to do to achieve the same result. One step on a mat is all that is required for the door to open. As for the other processes, many do not even realise that something is happening. The job gets done without most people knowing, performed by different parts and components attached to the door that could communicate together to perform a series of tasks as planned. They do it well and repeatedly.

That are so many other solutions developed using automation. The principle is the same. Get something else to do the work and organise the sequence the way you want it to work. Everything that works requiring no or limited input; was consciously and deliberately made to function that way. There is something hidden responsible for the actions, series of activities, and visible order we enjoy. The truth is that something is going on behind the scenes responsible for what we see.

CHAPTER 4
Learn from History

There has always been a need for human beings to move from one point to another. It could be for various reasons like work, social interaction, exploring the environment, expanding influence over other areas, or getting what is required to survive. How far can you go, and how could you move from your starting point to your desired destination?

It is possible to move from one point to another, although one may encounter different challenges on the way. The geography of the earth and its features help us appreciate the need for diverse solutions to facilitate movement. It has brought the need to explore all the possibilities available for moving people, animals or goods. The possibility of moving on land, water and air became the basis for the solutions that have improved transportation. They represent the different options that have evolved over the years.[1] The world has witnessed the emergence of all kinds of innovations. They have transformed our ability to move on land, on water and in the air.

The first known option is the use of the legs. Do it all by yourself by crawling, walking, hopping, jumping, and running to move from one point to another. Can you imagine walking from London to Scotland? As people moved from one place to another; some limitations made it difficult, if not impossible, to complete their journeys. They began to take advantage of animals like donkeys, horses and camels for transportation from 4000 BC to 3000 BC. Their involvement in the process was reduced, which made the experience easier and faster. It proved helpful when it involved very longs journeys. Life became a little easier as they only had to sit on the animal and instruct in the desired direction. It was enjoyed and celebrated, but the limitations soon became apparent when more people had to travel or when heavy loads require relocation from one point to another.

The invention of the wheel in 3500 BC made a solution to the problem a possibility. It brought about the advent of the cart. The same animals used for transportation were the source of power, but now they could transport more people and more load with less involvement. There are so many inventions that have helped the movement on land. They have also reduced the input required from the individuals. They include solutions with one wheel like unicycle to solutions with two or more wheels like bicycles, motorcycles, cars, buses, trucks, trailers, or lorries.

The invention of the engines was an improvement that had an impact. Its introduction and incorporation to transportation solutions further reduced the involvement of human beings. It became a reality to move one or more individuals from one point to another, including goods with only one driver required. It was a welcome solution that helped to reduce the energy dissipated by the individual involved.

It is interesting to know that the number of passenger cars that travel on the streets and roads of the world today is estimated to be over 1.4 billion, with about threescore and ten million new ones produced every year.[2] Every country in the world is taking advantage of this technology and using it to fulfil its goals.

You can start the car with a key and move following the manufacturer's instructions. You do not need to know what is going on in the engine. It just works. Cars have Engines that can run without so much energy dissipation from the driver. Some have automated gear systems which does more of the work. It has helped so many with a disability to move from one place to another on their own. There is no need to worry about changing gears and depressing the clutch every so often, as it is all automated for you so that you can conserve the energy and focus on something else.

Cruise control is another automation feature in cars. It helps you travel at a set or desired speed, and the car keeps moving at that speed without the need to do anything. You can rest your feet. It has helped many people stay within the required speed limits to avoid fines in places where you are issued a penalty charge for over speeding. It has helped some others to improve efficiency and reduced fuel consumption.

We have now seen the emergence of the driverless car.[3] It offers the opportunity to enjoy the benefits of automation. The decisions are made for you while you sit in the car, and you benefit from the technology as you move from one point to another as scheduled. The involvement required began to diminish with automation, and gradually it is almost zero, yet achieving the same result or even better.

Trains also have a similar history. The source of power was either by animals or humans before the 19th century. The invention and incorporation of the steam locomotive to trains made it possible to work with reduced involvement. Rail transport has also witnessed improvement and further transformations using electric power, diesel power and other technologies to reduce human engagement. Today we have high-speed trains that are faster than cars. There are trains with a speed of up to 267mile per hour, like the Shanghai Maglev, which uses magnetic levitation (maglev) technology.[4]

In London, like so many other major cities in the world, there is the Driverless Light Rail, a commercial train that can operate without any staff on board. It fulfils the same aim of transporting passengers, but the human engagement required is close to zero. It works. As long as you can get on board and alight from the train, you can use it with no staff required. It can be of great help to many countries, solving transportation challenges without added expenses—like flying in a professional or expatriate to be driving the train. You can get what you need to solve the problem you have without unnecessary extras or expenses.

Water covers about seventy-one per cent of the earth surfaces. It means that if you desire to travel far and wide across the earth, you are unlikely to be successful, except you can move on land and also in water or air. There are so many places on the earth surface that you cannot travel to on land alone. A body of water will prevent you from fulfilling that dream of arriving at your destination. You could either learn to swim across or engage the help of a means of transport that could move in water and carry your weight. It was this need that brought about the idea of boats.

The world's oldest discovered boat is a 3-meter-long Pesse canoe constructed from the hollowed tree trunk of a Pinus sylvestris around 8,000 BC.[5] Tree trunk? Yes, as long as the material can float in water when joined together, it makes a raft and serves as a boat. Bamboo, wood logs and reeds are examples of some of the raw materials that have been tied together with vines or palm fibres to solve the problem of moving on water. These early rafts, among other uses, served as fishing platforms, allowed transportation across bodies of water and even formed floating islands for villages. Driving or pushing the raft forward is achieved by pushing with poles, pulling with ropes or paddling.

The hollowed trunks reduced the involvement of the individual moving on the water. Swimming will surely take more effort and time. It is also a skill that needs learning. It implies that only individuals that can swim will be able to cross. The boat is a simple solution that made it possible to move on water and removing the limitation that was once a hindrance to movement.

Boats and water transportation have transformed over the years, as engines replaced and reduced the human energy required. Paddling is no longer necessary because something else is doing all that work, and the boats can move as desired. There are significant changes in water transport as we see the use of better raw materials, producing bigger, faster and more efficient means of water transportation, which has all reduced the involvement for individuals that desire to move on water.

We have now seen the advent of an 18-deck cruise ship, which is currently the largest passenger ship, measuring 1,188 feet, called the Symphony of the Seas. You will need to add another 53 meters to the tallest building in the United Kingdom and the European Union, the Shard, to get its length. It can carry 6,780 guests and 2,100 crew.[6] Can you imagine the human effort that can achieve this feat in those days? A total of 8,880 people on the water; relaxed, doing different things while the ship is moving. They could be eating, drinking, reading, watching movies, sleeping, and doing any other activity of choice while on the cruise. If automation were not in place, can you imagine that as a possibility? If it were possible to have the same number of people on just logs, almost everyone on board would be paddling all the way through. There will be no time or breathing space to attempt anything else.

Paddling of the boat was more engaging as it involved the continuous, repetitive activity of almost everyone on it. Subsequent improvements have made the involvement of the people reduced to insignificant. If compared to what it would have been in the past, there is a different experience, as they could travel on water and use their time and energy to do something more important while they still achieve the same goal of moving from one point to another. There is still propulsion, but in modern cruise ships, the human effort producing the power and some other human involvement has found a suitable replacement in systems. The power source for the propulsion and the operations are from a gas turbine or diesel-electric engines. It has reduced the involvement so much that only one helmsman controls the ship. Better results can indeed become possible when we learn to do more with less of our engagement.

Movement in the air is another way of transportation, first demonstrated by animals as they flew from one place to another of varying heights and different distances. The animals that fly have different shapes, sizes and weights, and they can fly and stay in the air for an amount of time. Some cover very short distances and heights, while some others cover longer. Birds, insects and only one type of mammal, the bat, are naturally capable of sustained flight.

The first flight technology to successfully carry humans is the hot air balloon as far back as the 18th century. The hot air balloon provided an opportunity for humans to move in the air from one place to another, which they could not naturally do. The hot air balloon relied on the technology that works on the fact that hot air rises. It was heated up using a burner, and anything attached to it moves up with it. It gave man the ability to move in the air with little involvement because the balloon floats upward as the air is heated up and slowly comes down as the air is allowed to cool.[7] It cannot be achieved naturally by humans. They cannot fly even with energy as they do not have a natural capability. The technology has taken the place of that limitation and made it possible with little or no involvement.

It is possible to move in the air as humans because we found a technology capable of producing the same experience that birds naturally have. We took advantage of it and can achieve the same result every time using the same technology without possessing the physical or natural ability unique to a bird.

The flight of the Wright brothers brought about a new dimension to air travel in December 1903. Their plane was powered and controlled, unlike previous attempts that had only one or the other. They were able to combine both, and this made it possible to create aircraft that are powered and heavier than air which eventually set a standard in aviation.

Aircraft have evolved ever since, and the world has seen all kinds of them, including bigger, faster and energy-efficient ones, moving humans, animals and goods from one point to another. The Airbus A380-800, a passenger plane made in France, could fly up to 853 people in the air. The aircraft considered to be the heaviest ever built is the Antonov An-225 Mriya, and the weight at take-off could be up to 640 tonnes or 640,000kg.[8] It is interesting to know that such a feat is achieved continuously with reduced involvement. It was not so in former years.

Can you imagine what it would have taken to achieve the same result even if humans could fly like birds? Thankfully, it is no longer necessary to think of that. A solution that works has made it possible to move in the air, and it has happened repeatedly time and time again. Humans fly in planes from one destination to another, enjoying the benefit of air travel and yet they still cannot fly in the air like birds. They can also achieve so much with so little.

How many people would have been able to get this done with their energy, even if they could fly? One trained flight pilot can successfully transport 853 people by air or move a cargo aircraft weighing up to 640,000kg from one location to another. One individual is sufficient because a suitable replacement has been found and implemented for almost everything. It works continuously and achieves the same results, so all that is left can be managed by one person.

A Monarch pilot, Sam Bray, was once asked an interesting question about the aircraft and its operations. I think the response is revealing and insightful.

Interviewer: 'How hard is it to fly a plane?'

Sam Bray: 'There's no point denying that the autopilot does most of the flight—it flies the aircraft smoothly and accurately. But with such a complex machine, we have to keep monitoring it to make sure it operates correctly. And in the unlikely event of any technical issues, we can let the autopilot fly the aircraft while we deal with the problem safely and effectively. On a regular flight, the autopilot does around 90 per cent of the flying.'[9]

A British Airways Captain, Steve Allright, also confessed that the plane could land successfully at an airport like Heathrow without the Pilot sighting the landing strip.[10]

What a revelation! It means that 853 people on board the aircraft fly in the air, across land and ocean. They arrive safely and on time at their desired destination with limited continuous human involvement because of automation.

What is doing all the work then? Some systems and processes help to achieve the feat. They are all connected with the various components performing their different functions as designed or programmed. They eventually made it possible to operate the aircraft with just one person, and that professional seems to have even more free time than we could have imagined.

The 853 people on board also have time to continue their life and engage in their desired activity, sleeping, watching, listening, writing, reading books, eating, drinking, using the restroom and many more of their choice. There are so many types of aircraft. They allow people to perform functions in the air, like in the home, office or professional functions, including attending meetings, medical care, military, or surveillance, among others.

How amazing it will be to free yourself from repetitive and complex tasks. It could be done effectively by a suitable replacement. You can focus and concentrate on more important things and yet still achieving the same or better results not just once but continuously.

Automation is not just for one field of human endeavour. Everyone living can take advantage of automation to recover and maximise their time for efficient and effective results. Automation is for all. If you need more time to do more things that are more important or you feel limited to do bigger things or affect more people, those are signs that you need to get something else to do it for you. If you want to do more or keep the results continuous and progressive, it is an important signal to watch out for, a reminder that you need to change as you can no longer cope.

Automation could be simple, and you can make it work for you no matter what you do. This simple definition may help you to see automation differently. Automation is a way, technique or method of achieving the same and continuous results with no or reduced repetitive involvement by using or replacing the engagement with anything capable of doing the same thing or achieving the same result effectively and continuously.

In other words, it is a solution that will help to get the same results, but most or all of the work is not done by you repeatedly. Simpler yet, it is making it work by itself or with limited interaction. Exactly! You can have your time back, you can have your life back to do more important things, and you would still get the results. It is just not done by you but for you.

CHAPTER 5

Opportunity

Many years ago, I was part of a team of six to support an organisation with over three hundred and fifty staff members. I tried to make myself useful, performed my role as expected, and tried to go beyond. I tried to find viable solutions to the diverse and emerging problems the staff members encountered in their daily work as they use the technologies we deploy and support.

I was gradually getting used to the environment. I implemented solutions that helped people do their jobs better by providing technologies that could improve their efficiency and performance.

Sometime later, I began to see the failure of components in the hardware and software. It could be the same parts in different devices or otherwise. They were old and obviously out of warranty. The problem began to grow, and the failures increased, which meant more people were affected. The loss of time for productive work also grew as many often complained about their inability to work as systems develop faults and shows in different symptoms that hinder their ability to wok and complete their tasks successfully or on time. I had to rebuild them or change the faulty parts or resolve any application related issues as the case may be to help the user get back to their work as soon as possible.

The issue soon became part of the discussion in the staff meetings. It began to gravitate to the top of the agenda. There is a problem that needs a proper and lasting solution, and we need to find that solution. It was not too long, after so many deliberations and suggestions, there was an agreement to replace them with new ones. The change will apply to the whole site of over three hundred and fifty members of staff.

The solution is here, some thought, but so many questions remained unanswered. There are more decisions to be made to achieve the desired result. How will it affect our work? How can we manage the change? Will the new device and software be compatible with other devices and software that we currently use? Can we afford it? Will it be worthwhile and represent value for money? Is it the best way forward? Do we have the skills required? Many other questions also needed answers.

It is often easier to suggest a solution than to implement it. It could be a good suggestion, for instance, to buy a piece of new equipment to replace a rusty one, but doing it may face more challenges than anticipated. This easy looking solution was not an exception. It encountered so many huddles that made it seem it will not be a reality after all.

Funding was a big one. Where will we get the funds? Will we have enough to complete the project? Every organisation has a policy on how things must be done. It restricts the organisation and determines what can actually de done. It also affects what you can buy and where you can buy it. It is not as easy as imagined. One could only imagine that it was that easy. An attempt to solve a problem has just created more problems to solve.

The attempt to solve the new problems progressed, and we found answers to some of the questions as we sought help as an organisation. The good news is that we got a promise for the devices, which was a great relief. We still needed funding for other parts of the project. Fortunately, we got that too. It came with a clause that the project must be over before the end of the financial year. Good news! We have the money, but not really because it is only ours if there is a possibility for completion before the end of the financial year. The answer to one question was necessary. Can we finish the project before the deadline to benefit from the funds?

Another race began to ascertain what the project entails. The possibility of completing the project in the stipulated time to benefit from the funds promised. There were so many options, but none put us in a comfortable situation to guarantee the promised funds. It was now about six months to the deadline given, and I began to think of a solution that could rescue the situation. There was a lot to cover for the project to be completed. 'What can I do?' 'How can I help?' I asked myself.

The challenge is now clear; it is a race against time. Whatever solution that can deliver the expected result in reduce time will solve the problem. That is when I began to think of how to get the job done in less time. There must be a way out. The so many problems we started with at the beginning are decreasing. At this time, we had fewer, and it was such that it was up to us to make it happen. If we could meet the terms, it was as good as done.

Have you ever been in such a position; that it is all up to you? If you can do more for less, the benefit is all yours? Are you struggling to keep a deadline that you know you may not likely meet if you keep working at the same rate? It is a great challenge, and it could be stressful. You have found the problem, you know what to do, but you seem not to have enough time to get it done. What would you do?

Every challenge comes with a great opportunity, and every difficult situation provides a platform for discovery and innovation. If you can find a solution to that problem, you will not only have solved it, but you will also have produced a solution that could help many others that have had, are having or will have the same problem in the future. There is nothing new, your current challenge was the problem someone else had yesterday, and it will be another person's problem tomorrow. Your solution is not just for you; it could help many other people that encounter a similar problem.

What will you do in such a case? There is a high risk. There is also a great opportunity. It is all hidden in a problem that needs a solution or a question that needs an answer. How can we probably change the devices of over three hundred and fifty members of staff with limited resources and limited time?

You might have heard people say; I wish I had more time, more hands, or more than twenty-four hours in a day. I can understand their plight. It is usually a comment made by people with a sense of frustration at their inability to complete their desired tasks or goal even though they know what to do and how to get it done. They cannot because the time they seem to have left will not be sufficient to complete the tasks. It could be so frustrating to lose such an opportunity just because of time.

The problem is simple. I need to do the same thing, to achieve the same result but in less time. It is a typical scenario many people face. If you can solve problems in less time, you can gain time for more effectiveness.

I discovered that there is one thing we had to do over and over again. It is a must for every device that will end up at the desk of each user. It is also one that will take a lot of time to complete for one device. It implies that it will even require more time for all. The task was to make it ready for every staff member to use. It was to install each device from scratch with a new operating system, including all relevant applications, customisations, and data transfer. It also involves testing to make sure everything works well and with other devices.

We had a small team with different areas of specialisation. There were limited resources in terms of tools, money, space and infrastructure. This project will not also exempt us from our usual daily work called business as usual or other projects that we might be doing. Emerging issues could often be an opportunity for automation. The thought was if we can find a way of reducing the time and effort, we would have found the solution to saving time. Whatever can help us save time in repeating the same thing three hundred and fifty times would surely help us.

As at this time, we could not get anyone to take on the project externally to deliver within the required time frame. The most crucial thing became time. The complete setup of one device from start to finish may take up to five hours or more on some occasions. How can we manage to accomplish up to five hours of work on each device and yet finish the project for three hundred and fifty devices in the desired time?

If the tasks take five hours to complete, it will require 5hrs x 350 devices. That is 1750 hours or 218.75 days or 10.94 months, assuming an 8-hour working day, 5 days a week, and 20 days a month as a typical working pattern. It will not achieve our goal.

If the tasks take 2.5hrs to complete, then it will require 2.5hrs x 350 devices. A total of 875 hours or 109.38 days or 5.47 months. It will also not give us the desired result. If we can reduce the time taken to make each device ready, there is hope that we can meet the deadline.

Some detailed calculation became necessary to know what we need and how to proceed. The details are as follows;

If the tasks take 1.5hrs to complete, then it will require 1.5hrs x 350 devices. That requires 525 hours or 65.63 days or 3.28 months.

If the tasks take 1hr to complete, it will require 1hr x 350 devices. It implies 350 hours or 43.75 days or 2.19 months.

If the tasks take 0.5hr to complete, it will require 0.5hr x 350 devices which will consume 175 hours or 21.88 days or 1.09 months.

If the task takes 0.25hr to complete, then it will require 0.25hr x 350 devices. We will need 87.5 hours or 10.94 days or 0.55 months.

It implies that if we can reduce the time taken to make a device ready from 5hrs to 0.25 hr, one person could make them ready in about 11 days.

It is just a part of the project, not all. All the other aspects of the project will need time to complete as well. The goal to reduce the time even further will be desirable for the project to finish as desired.

There was so much to do, I could not solve all the problems, but there was something I could do to contribute to make it a reality. There is one major challenge that was clear, and that was time. If we had everything required, there is still that crucial factor that can either make or mar the project, and that is the ability to get it done in time.

I began to look at the problems before us. I identified the ones I could help with and focused on them. They were time-consuming and repetitive tasks, and my focus was on reducing the time it will take to complete those tasks on each device.

I took on the challenge and began to work on the possible ways to solve the big problem that could help us succeed in the project. It provided me with the opportunity to seek a solution, which could help us achieve our desired goal.

CHAPTER 6
One-Touch

It was the start of another day, and it was getting closer than when we started. There is now pressure as the desire grew to seek concrete evidence that the project will complete. It is one thing to plan but another to do. Time continued to tick, and it was time to make things work.

I have identified a problem that I could commit to solving and began to work on it. I did a series of calculations, and I had the idea of the desired time required for the project to succeed. If we could achieve close to that time, it would be helpful. More questions arose that needed answers. 'Is it possible? Can it be done? Do we have the tools? Has it ever been done? Will it work? Will it be acceptable? Will it comply with standards? Will it be a worthwhile venture?' There were so many questions, some of which had the potential to discourage and frustrate the project.

Some people have discontinued their pursuit to solve a problem because of some or all these questions. They neglect the idea. They are drained of strength to carry on, but I would like to encourage anyone in that position that if an idea or solution has ever come to your mind, you have the potential to bring it to fruition. Do not give up, but continue to the end.

The more I progressed, the more confident I became. I saw more possibilities and found new solutions. I was able to see that it was possible to complete the tasks even faster than I earlier anticipated. I had to spend more time searching for solutions, but if it works, it will pay off in the end because it will mean that we will complete the project on time.

At a constant rate of 0.17hr to complete tasks on each device, it will require 0.17hr x

350 devices. It implies 59.5 hours or 7.44 days or 0.37 months. Only 10.2 minutes. That will be brilliant. It will get the job done. The problem has brought an opportunity, and I began to search for ways to make it a success.

The challenge I set to reduce the time taken now needs action. Completing the activities for each device in the shortest time possible will be the game-changer. I took on the challenge, and I began to list all the tasks required and search for ways to accomplish the same with reduced time. I sought out the possible command-line equivalent for each one and the corresponding environment for it to work.

I dedicated so much time to this and found so many options that could make it possible to perform each task without my involvement. I then had to test all of them to ensure they worked and can do what was required. The last part was to combine all the individual tasks, ensuring that they can all follow the desired sequence, checking that each completes before the next could run. There were so many challenges combining it all into one, so I had to try different options and alternatives until I got something suitable.

You may not appreciate your ability until you are challenged or faced with a difficulty that causes those inactive abilities to come alive. The world enjoys the benefits of innovation today because someone saw a problem and decided to fix it.

I began to work on all the tasks required to make the device ready for each user. There were quite a lot of them, and many of them required user interaction. You are required to answer a Yes/No question or agree to an agreement and confirm a change or consent to proceed with a required action or restart the device for changes to take effect or respond to an error message that needs a decision. The process literarily comes to a pause until the expected action takes place.

You may probably be familiar with the installation of an operating system on a computer. For instance, you decided to multitask—doing many things at once, you started the process by booting up the computer from the installation media. The process begins; it seems it is doing something that does not need your attention and that it will surely be a long while before completion. You thought, 'I can probably use my time wisely by engaging in some more productive activity.' You chose to engage in some other productive activity instead of sitting or standing still staring at a computer that knows what it's doing anyway. 'Good use of my time,' you say, 'and at least by the time I return to my computer it would have completed the installation, and I would also have gained time doing something else, so it is a win-win situation, I would have killed two birds with a stone.'

If you are smiling or laughing, you must have crossed this road before and found yourself in this position. Yes, I think so. You arrive expecting to see a completed installation of the software. Not really, you saw an unexpected screen message instead; 'Please press Enter to continue' 'Choose the desired option and press Enter', 'Enter Yes or No', 'Are you sure you want to install this program? Press Yes or No to confirm' There are many other similar interruptions and messages depending on the software. The prompts usually bring the installation to a pause which requires your interaction to progress. It may have appeared some few seconds after you decided to leave or some minutes after it commenced. 'Wasted my time' is one of the likely responses heard while staring at such on the screen. It would have been great without the interruptions. I thought of that too. You could then actually save time. The idea was to make it run from beginning to end, doing everything without any interaction. If I could achieve that, then it could help us achieve our goal. It would simply work by itself.

It was a good idea, but a solution to all of these interruptions was crucial to achieving it. It implies that for each device, the software installation should complete with no user interaction. It is not just the operating system, but all programs, customisations and configurations as well. It must also happen within the specified time.

Day after day, I sought solutions to every process and task. I searched for ways to incorporate all the actions with no interaction. I was able to bring everything together in one single operation, but not without issues. If it works as expected, the device should be ready for use after the specified time.

I presented the solution to my colleagues regularly for criticism and also to provide updates and progress report. I got encouraging feedback, some of which had to push me back to find some more solutions to solve the concerns raised.

Many times, we see criticism as bad. It was through feedback that we saw some things that needed improvement in the project. They were not frustrating me, but they were my helpers. I needed their input to improve on the project and produce better work. I may never have seen it necessary to go in search of more solution without them. The product is for the people, and if it does not meet their needs, we have not fulfiled our goal. The satisfaction of the users becomes our fulfilment.

I wrote down each comment, praise and criticism. I took it away, worked on it and returned with solutions. I got some more, handled them the same way until there was no more. The solution answered all concerns and questions.

Your product is as good as the level of criticism that it can endure. Do well to appreciate feedback, whether good or bad and improve your product. It is a reflection of the groups of people that will use your product. If only ignorant people appreciate your work, it could imply that your products will be valuable to those class of people majorly. On the other hand, if it endures criticisms of the knowledgeable, it will appeal to that category of people too.

You cannot please everyone. A product may not satisfy everyone either. It must be suitable for those that need it. You can improve your products in line with your goals and the needs of those that will use them.

It was encouraging to know that it was a solution fit for the purpose. I finally packaged it in a portable bootable media, with all the operations integrated to function as one. You can boot from the media and walk away. You return in about 10 minutes to a ready custom-built system. It completed in just over 9 minutes. It made it possible to complete the installation of all the devices in 7 days by one person.

Problems do not necessarily need to be solved by your two hands repeatedly; you can also engage the relevant help and capable replacement to get the job done. You can then get even more job done.

It became the One-touch solution that saved the day. All that was required was that touch to get the process started. It got the job done.

CHAPTER 7
The Value

The value of a solution is a function of the problem it solves. Many critics will often say that everything is easy to do. They could say things like; 'It is nothing, they could do it, it is not anything extraordinary, it is not such a big deal, after all, even my five-year-old boy can do that, that is just common sense, that is nothing, I can do that better, I have done more than that...', to mention a few. Yes, they are right. There is nothing new, but using easy and old ideas to solve new and current problems is essential and required. It is what makes any solution relevant, thus making it valuable. It makes it seems new.

You may know many things but may not add value to anyone with what you claim to know. You cannot connect the knowledge to the need to help solve the current problems and challenges. No one can benefit from what you know, even though it is simple, known and old. You are not valued because the world only values people that contribute to their goals and help solve their difficulties and challenges. The idea of automation is not a display of expertise but rather a focus on solving current and future problems by adding value. The value is in the problem it solves or the solution it provides.

There will always be a better method or process to achieve any given result. Ten people that gather together can solve the same problem in ten different ways. That is why we can have many restaurants, for instance, serving different types of food. You will observe that the kinds of customers are not the same. Some could target office workers, while another focus on students. One may be established for parties or celebrations, while another casual diner.

There are several types found in diverse locations, providing their services at varying times of the day. One could be in the village, providing a family-style dinner for special occasions, while another in the city providing early morning breakfast to commuters going to work. They differ in so many other things too but solve the same problem. They all provide food, but there are many of them, and they achieve the same purpose of feeding people in so many different ways.

There will always be a need for a solution, and it could be improving on an already existing one. There is nothing new. If you identify the problem around you or an existing solution that could benefit from enhancement, then that is another solution. There were so many restaurants before the advent of fast-food restaurants, but the fast-food restaurants solved the problem differently or improved on an existing solution of feeding people. They thought of a solution to feed people on the move. It was not that people did not know where to eat, but they could gain time if the food were on their way as they travelled on the road.

Some time ago, on one of our holiday trips to Europe. We stopped and visited one of the stores in Germany. One of the things I discovered was that the store had a self-service bread machine. If that is so complicated and technical for you, let me try again. It means that you can dispense freshly baked bread and pastries by the touch of a button. Yes, imagine the value. You walk towards this bread machine and read the beautifully presented information displayed. You choose from a range of variety and confirm by depressing a button. Yes, the selected option will be prepared and dispensed as if it just came out of the oven. You will not see anyone. You need not talk to anyone. Is that not value? Amazing! It is a solution that provides a unique experience for customers using automation.

The cost of a tea bag is about two pence, but the value of tea made out of the same tea bag is different depending on the problem it solves and how it is processed and presented. Customers that need a cup of tea pay varyingly for it in different situations or locations. Some may purchase at £1 in a local shop, £1.50 in a self-service tea or coffee vending machine, £2 in a restaurant, £5 in a hotel and more in a luxury accommodation alongside other services. It implies the value you can make out if it depends on its use and the problem the material is trying to solve. The most expensive cup of tea is likely to be one sold in a hotel. One offers it for £165 a cup or about £500 for the whole pot of tea.[1] The value increases by the processing and presentation of the raw material.

There is a possibility of improving on every solved problem. The improvement to the solution could add a new value that could make a significant difference. It could be comfort, flexibility, options, privacy, simplicity, or uniqueness. The difference could go a long way in uniquely helping more people.

In most countries, parking a car could be a challenge around town or city centres, but there is already a parking solution. Many still turn up and cannot get space. A solution could be to book ahead, managed using an automated online booking platform, with reduced engagement. Another one is to encourage everyone with unused parking spaces around their houses or properties to let it out for an agreed time. The available slots can be seen and booked ahead, using a relevant online platform without extra involvement. You can increase the value by using it to solve a problem.

Healthcare could be expensive in some countries and not available at all in some others. A solution could be to make it affordable and possible when it is needed. You can connect medical experts to patients using an automated online platform. It will help save money on transportation, property management, and other running costs. The patient will be able to see and book appointments as desired. The specialist can also publish and manage availability throughout the year. Cheaper fees with more patients will end up bringing more income than expensive with few patients. It could work for professionals with other commitments like family. The flexible arrangement will help them to do their job without neglecting other areas of their lives.

It is a difficult task to get genuine car parts in some countries. One can attempt to solve that problem. You can set up an automated platform to register trusted dealers and also verified buyers. It would connect the dealers to prospective buyers through the online portal, and the dealer offers the same guarantee with a promise to return or refund. The solution will continue to run with reduced involvement, but it would continue to solve the problem.

There will always be a need for training and development. Many organisations are willing to invest in it. Training and courses could be expensive in many places, and some cannot afford the cost, especially when international travel is involved. The solution could be to provide an automated platform online, connecting the specialist with the students. It will be saving the time required for transportation, the cost of a classroom, and all other costs like flight tickets, feeding and accommodation for the trainer.

Many support companies provide a twenty-four-hour service, using a model that follows the sun. It is a concept that utilises a pool of global human resources located in different time zones to provide continuous service. The service is always available but serviced by people from various locations. Everyone works usual hours in their time zone. An automated platform could manage the employees as well as the clients.

Some other organisations provide services based on an on-call basis. An automated solution can also help to reduce involvement. All trained personnel registered with an agreement to provide a service will be called or alerted by the system. They could pick up the call to complete the requested task and get paid for their time from anywhere around the world.

A lot of people gather together for conferences, meetings, and collaborations. They spend time and money on transportation and other expenses which could be avoided by using automated systems.

Online meetings are another solution to the same problem. It may not work for everyone at every time, but it a solution that could transform a life, community, university or nation.

Every vehicle owner attends the garage to service their automobile. That is a solution, but the Mechanic sits there some of the days with nothing to do. An automated platform could help solve the problem, such that people can book ahead for the service to match the time the Mechanic will be available to provide that service.

A local company providing insurance services can reach more people around the nation for less. It could benefit from an automated online web portal, which could provide insurance quotes for customers at any time of the day. It will save money on office rent, maintenance and other related costs.

Imagine a local artist whose work is not attracting the desired attention in the area. It could be uploaded online, made available to the global market. The images will connect to lovers of the work around the world.

A lot of people have skills that are in demand around the world. A portal connecting the trainer and the trainee will be beneficial. It is an opportunity to train, mentor or advice people. The skills could benefit the global market and transform someone's life. Many people have skills that are not appreciated or oversaturated in their environment. They are in short supply in so many other places globally. A solution that could connect people in this manner will be a brilliant idea.

Our experiences are precious. Many have gone through so much that their experience could help many other people around the world. They may seem irrelevant in their local environment, but they are needed somewhere else. How can they achieve that? An online portal that could connect different life experiences and people with helpful information would be a good idea. Allergies, Uncommon diseases, life experiences, phases of life, and many more could be areas of interest.

Some small businesses or stores stock up their shelf so that people could walk in, find what they want, then buy. It is a good idea, but you have to travel to the store, check, and probably return without buying anything because you cannot find what you like sometimes. You have already incurred the cost of transportation, and time is also lost. It could be made available online with pictures, detailed information and availability. Customers could either order online for delivery or attend the store to pick up depending on their choice.

Imagine a small business owner selling children clothes. The cost of incurring a shop and a shop attendant could be high, thus affecting profit. The business can be made known to more people for less. A suitable online platform can help. A simple website with detailed information will make it easy for customers to view and make transactions. It will be available twenty-four hours a day. It could save on the cost of the shop. It could also increase sales if both solutions are combined.

Everyone needs a solution. Many problems need one. Identify the ones you can solve. There are many around you.

Automation will be useless if it does not solve any problem. It will also be worthless. The value comes if it could help recover valuable time, energy, and resources.

Think of how you could reduce your input. Can you reduce the energy, time invested in your desired activities by 50% and still get the same result? If yes, that is value. If you can even reduce it further, using only a tenth, that will be brilliant. You have recovered time and energy that you could then use to help others or devote to important things you cannot automate.

◆◆◆◆❖◆◆◆◆

The value of a solution

is a function of the problem

it solves.

◆◆◆◆❖◆◆◆◆

CHAPTER 8
Think Automation

Do you want to profit and be fulfilled doing more of what you need to do with less? There is a way out. Learn how to solve your problems by using something outside of you, to free up eighty per cent of your time. If you have achieved that, you can help others free up their time. There are opportunities everywhere. Anywhere you see repetition—doing the same thing again and again, think, there may be a brilliant opportunity waiting for an innovator.

Search for nations, companies, communities, groups, professions, and people. Think of how you could solve their problems, reduce their time, reduce their dependencies, improve their efficiency, accuracy, and you will become relevant to their progress, and we all soon make the world a better place. Many human beings spend most of their time doing things that animals can do. They expend their energy and time on activities that could be carried out effectively by a capable replacement. We can automate to free our time and thus invest our high-value system in doing what cannot be automated.

If you see any process anywhere that is done the same way, again and again, using human beings, there is a possibility that a change could free those high-value individuals to do something of more value. Critical thought is the starting point for the change that will bring such transformation.

Scenarios

Let us consider some of the solutions we see around the world.

Amazon provides services for individuals and corporate bodies around the world. Many have benefited from these services without any physical contact with anyone in the corporation. They have been able to add value using concepts that have enabled them to reach more with less. They provide infrastructure, deploy enterprise-class solutions and various services automatically.

They have been able to figure out what many have been doing the same way for many years. They made it possible with less involvement. You can buy a book, read a book, or publish a book on Amazon from the comfort of your house. It works at any of the twenty-four hours of the day. Thousands, millions or billions of people from different parts of the world can also do the same while the owner is busy doing something else with his time. The number of employees that would have manually attended to all those repetitive requests will be multitudinous.

Uber, an American company that offers global transportation services, excels based on the same principle. They connect travellers with car owners with the use of application software. They do not own the cars, nor know the travellers, nor own the car park, nor know or visited all the locations, but they found a solution to what people do all the time without the many people involved in doing it. They may never have been to visit many of the locations where they provide services. The business keeps running while the few people that run the corporation can engage their time in something else.

The American multinational e-commerce corporation; eBay is another example. You may now begin to see how they also have made automation work to their advantage. Some people want to buy; others want to sell. It has been happening the same way for centuries. The corporation created a platform to connect the buyer to the seller. They left it to work while they can spend their time doing something else. Can you imagine the human involvement and many other things that would have managed such a global market place if it were physical and manually run? I hope it is becoming clear to you.

The human involvement in running the market place has reduced to the barest minimum. The solution eradicated so many other things like storage and costs incurred in a traditional setting. Many more opportunities exist to connect buyers with sellers or clients with service providers. Set up the platform and let it do the work while you are free to do something else.

Airbnb is an American company that provides vacation rental through an online marketplace. They connect renters with landlords and allow the platform to manage the relationship while they do something else. They could be making money while sleeping because they have created a solution that works with little or no involvement. They do not need to build houses, nor are they required to buy them, nor do they need to know the people, nor do they need to be awake to see what is going on. It works whether or not they are there.

Some ideas

The general principle of all the above is simple. Look for the problems and challenges of people, and try to solve them through automation.

You may not own the asset or product, but you could connect those that have it to those that need it. You can then enjoy the benefits of your innovation.

Online Car park reservations

It will help people looking for car parks anywhere in your country or any country you have access to, or anywhere. All this required is to connect through a website to make or manage the reservations. The people that need parking and those that could provide the parking in that location can conduct transactions anywhere in the world.

Tourist Holiday Package

Many people travel to different countries and wish to know more about the country. If you are familiar with the nation or have access to people who do, the lovely attractions and landmarks that may be of interest to tourists, you have a treasure, and you can solve a problem.

You can get a list of all attractions, create a package for a week that will take tourist to some attractions for that week, and return them to the airport. Put that online, and everyone around the world can see it. You connect the holidaymakers to a local attraction. You can now create packages with varying durations, for different interests, for various age groups, and so on. When it is up and running you will get your reward. The same model could help provide other services for accommodation, restaurant, groceries, snacks, fast food, education, or whatever is a need of human beings.

You also reward others that are part of the process. The joy, excitement and appreciation from a tourist with a positive experience are priceless. It could be a local business that suddenly gets a rise in sales or a group of locals that gets employment because of the solution. It has brought a positive change and would be valued and appreciated.

A colleague of mine once told me of his experience when he travelled out of the country on vacation. He lodged in a local hotel and enjoyed the local services. He was impressed by a waiter and gave a tip for the lovely and honourable service. In his country of residence, it is such an insignificant amount of money, so he could afford to lose it. It was not the same for the recipient because of the exchange rate. It was significant to the recipient. Guess what? That same amount of money to the waiter was more than a weekly wage. You can probably imagine the joy and fulfilment. It is because the global market was made available to the benefit of a local skill. The local business will also profit with the possibility of expansion, thus spreading the gain to others.

Tourists travel globally, and many more want to travel to where they have never been before, but there are so many questions and fears. There could be concerns about safety, where to stay that will be safe, what to eat that will be good and decent, where to find the landmarks and attractions, and so on. These unanswered questions will discourage many tourists from visiting such places. Some will like to buy some souvenirs, know what to do for fun while there, and many more.

Can you solve these problems and many more? If yes, advertise your services and ensure that it is safe and reputable. You can automate the process to connect the tourists to the service providers. You can also create a package by working with all the trusted service providers. You can negotiate for all the services for a trip and ensure they are available at one point of contact or what some call the one-stop-shop, a term that means getting everything you need through one point of contact. You will be surprised at how people need what you have from all around the world.

Look around you and observe. Is there anything people do over and over again? Can you figure out what companies will always request repeatedly? Have you noticed an action done at a regular frequency? Have you seen professionals at work, and all they do is respond to the same type of questions all day long? Have you visited an office that provides the same information to everyone? Think! You may be passing by opportunities to make a difference without knowing.

Is there something you are doing that is taking too much of your time, but it is all repetitive? There could be a way to reduce your involvement and still get the same result. Think! Providing the solution to problems will always yield dividends. If you keep doing the same thing over and over again and expecting the same result: Think automation.

◆◆◆◆❖◆◆◆◆

If you keep doing the same thing over and over again and expecting the same result: Think automation.

◆◆◆◆❖◆◆◆◆

CHAPTER 9
Creativity

Some years ago, I needed to book my car for a test. The Ministry of Transport test, called MOT, is an annual vehicle safety test carried out by mechanics. They are required in the UK, like other countries, for most vehicles older than three years.

Many local mechanics are qualified to carry out the test, but many could be frustrated and never profit because of their business approach. Let us look at an automation idea that could help a local mechanic or chain of test centres improve productivity using automation. It is also applicable to many other business ventures.

Anytime I need such a service, I usually look for deals that represent the best value for money. Yes, I got a good deal for half the standard price. Asides from the money I saved, I got an outstanding concept. I discovered something interesting that I will like to share with you.

This company has a website to book for the same MOT. You can book the MOT even outside the working hours of 9 am to 6 pm. It helps many more customers not available within the opening hours to book. That is one advantage. The price was reduced, which will be an attraction to so many others. The booking process is online. You do not need any staff to manage it, thus saving money on human resources. Payment is online and confirmation sent by email. You can attend the MOT with a transaction reference or a print out of the email.

Some more benefit that I saw. The MOT will require mechanics or testers to accomplish. Let us assume for argument sake that they are hired on a permanent contract to work eight hours every day. They may not be fully engaged. They cannot guarantee that customers would turn up for all eight working hours every day. They may be idle for the first two hours of the day and serve a couple of customers for the next two hours. They could have many customers for the next three hours, of which they can only attend to five. The others will probably go somewhere else or try another time because of the waiting time or many queues. In the last hour, one car arrives and just about the time to close few cars turn up, but they have to turn them away as the cannot finish the work before they close, and they will probably suggest another day.

They were able to complete eight cars for the day, but it seems to be below their capacity. If it takes half an hour per car, it implies sixteen cars in eight hours of work, not eight. It could increase if there is a facility that could make two mechanics work in parallel. It will double to 32 cars for two mechanics, 48 automobiles for three, and 48 vehicles for four mechanics.

The example shows us that the same test centre could have completed 16 cars instead of 8, using only one skilled staff. It operated at an efficiency of 50%, an income of 50%, and a value of 50%. I found that they could double income, efficiency, value, and customer satisfaction, even with a price reduction. Applying concepts to produce more with less will help people to book for all the available times far ahead.

You may then argue that if there is a 50% reduction in the price and they served double the number of cars, they will still have the same amount of money. So 8 x £50=£400 for the day per skilled staff, and at half-price of £25, implies 16 cars x £25 =£400, so what is the point, no gain. Here is where the profit lies. They could serve more customers, which is an advantage. There is financial gain because the discount was varied to attract customers to quiet times of day: as shown in their record or history.

Let us go back to my earlier example of a typical day at the workshop. The company considered the time of day when giving the reduction to fill all time slots. The first discount of 50% is for the time slots in the first couple of hours. They are usually free of bookings. They decided to give less reduction afterwards as they have some customers that turn up at that time as well. They offered the service with a 25% reduction, to attract some customers to book the time slots in the next two hours. The customers could then take advantage of the 50% discount or the 25% discount depending on their availability.

There will be competition for this time, and the limited time slots will get filled quickly. Many will keep looking for those days in the future. The test centre will have all their idle times booked far ahead, and thus there is a guaranteed income for those times. The process is managed automatically without extra involvement.

The 3 hours that follow are the usual busy times when many are willing to attend. It may represent when most people are available or because of the work people do. For instance, so many businesses have busy times and quiet times. A restaurant could be busy between 12:00 and 14:00 if it is close to a working environment where the workers have a standard break time within that time range. They could also be busy just before the usual work start time or finish time as people may want to eat before they go to work, during lunch or after work. All other times could benefit from a discount that would encourage other people that do not have the same work pattern or are flexible. This understanding can help the business distribute the sales to maximize all times and serve the community better.

These 3 hours also attract a discount. It will encourage people to book, and the test centre could also compete fairly with others. Online booking gets a 10% discount at these time slots because they usually have more customer than they could handle anyway. It is also attractive to those available at these time slots as the discount, even though not as much as others, is still cheaper than other providers of the same service.

The last hour is also not busy, which could also benefit from a discount of 50%. This distribution will ensure that all available timeslots get filled up. The times for breaks and other unworkable times will not be available to book.

It means that for one skilled staff employed to attend to 16 cars, an income of £570 will accrue. The breakdown is as follows; the first 2 hours will be £100 (25x4), the next 2 hours will be £150 (37.5 x4), the next 3 hours will produce an income of £270 (45x6), the last hour will bring in £50 (25x2), which give us the total of £570. It represents a 42.5% increase in income per mechanic, so if the test centre has a facility for four more mechanics, they could have 170 x 5 extra daily. It translates to £850 extra income every day by using what they already have. That could even pay for the daily wage of the mechanic times over.

You could do more and achieve more with less of your time and involvement so you can focus on many other crucial things. People can book your service anytime and anywhere. They don't even need to know you or see you, they may not even like you, but if you have a solution to their problem, they will dive in. That is the power of automation. Are you still wondering which time slot I booked for my MOT and how much I paid? You are right! I chose the cheapest one.

If you employ a staff, to make bookings, cancel and amend bookings for MOT just for you to have people attend your MOT test entre. The individual will keep doing the same thing every day to achieve the same result of booking people for all the available time slots. Why not think of a change that will free up the time to do what cannot be automated.

Many people are engaged in activities that do not help make the most of their abilities. This example shows that technology can help to fulfil that role and yet deliver results. If you keep doing the same thing and you expect the same result: Think! If you get laid off because of the implementation of automated solutions, it is only proof that you are worth much more than what you were doing and you could do much more to get more result if you harness your gifts and skills.

Imagine a typical day in this role. People will either call or turn up at the test centre, and the conversation that ensues will usually follow the same pattern like the one below. The details are for illustration purposes only.

Customer calls AB Test Centre.

Staff: Good morning, AB Test Centre; how may I help you?

Customer: Hi there, I would like to book an appointment for my car.

Staff: Is that for MOT or service?

Customer: MOT, please.

Staff: Ok, sure, I can help with that. One minute, please. Let me check my diary.

Customer: Thank you.

Staff: What day would you like?

Customer: Today at 6 pm, please.

Staff: I'm afraid the slot is not available.

Customer: Ok. What days are available?

Staff: Monday 7th, or Tuesday 8th at 4:00 pm, Will that work for you?

Customer: Not really.

Staff: Let me know when you are available, and I could check that for you?

Customer: Emm. On Monday, I have an appointment with the doctor at noon, and Wednesday I will be going to the gym. Friday will be better at 5:00 pm.

Staff: Ok. I have some free time slots at 4:00, 4:30 and 5:15. Which one would you prefer?

Customer: I would prefer 4:30 pm if 5:00 pm is not available.

Staff: That is perfect. I will go ahead and book you in.

Customer: Thank you.

Staff: Can it take your car registration, please?

Customer: Yes, AB65WYZ

Staff: Is that Honda civic?

Customer: Yes, please.

Staff: Can I take your name and address, please?

Customer: Yes, My name is Jon, and my address is 10 Alphabet Lane, London. PN12 9LL

Staff: Thank you. It is fifty pounds, please. Are you paying with a card?

Customer: Yes, Credit card.

Staff: That is fine. Can I take the long number of the card, please?

Customer: Yes, 1-2-3-4-5-6-7-8-9-0-1-2-3-4-5-6

Staff: Thank you and the expiry date?

Customer: It is January 2030

Staff: Can I take the name as it appears on the card?

Customer: Jon Bil

Staff: Can I take the card security code, please?

Customer: 4-0-4

Staff: Thank you. I will process that for you now.

Customer: Ok.

Staff: It is all booked for you now.

Customer: Thank you.

Staff: Looking forward to seeing you on Friday at 4:30 pm

Customer: Thanks for your help.

Staff: Bye.

Customer: Bye.

I wonder what your thoughts were as you read through the conversation. Can you imagine doing it every single day? The information required from the various customers will always be the same. There are so many solutions that could help reduce the involvement and still produce the same result. It is achievable by a phone system; programmed to ask, receive and process the required information. There are so many companies that go through the same process every day. They do the same thing every day to book cars for MOT or other services and the result expected is the same. There are other solutions like the idea of the website mentioned earlier. Everyone can input the required information, check availability, make the payments, get confirmation and so on while the mechanics are busy doing their work. They only have to worry about confirmed bookings for each day and prepare to receive them.

The solution is set up once but continues to produce the required result with limited involvement. It is an advantage to get the same output but allowing it to work without you. Some other use cases may benefit more from the idea, where a series of processes complete without any intervention.

A training company could offer their programme through a website. The customers can purchase the item of their choice, get a receipt and access details by email. They can then log on to a secure area, using the details to download the materials or watch the videos. Amazing! It is a complete process, yet without seeing anyone. It continues to run even when the staff or trainer is asleep. The customers continue to buy, whether it is day or night. It is now independent. They invested time to make and upload the materials. They did it once and left it to continue to deliver.

I hope you can relate to this truth and find a suitable application that will make a difference to you and the world around you. You have enjoyed some of these services; it is time for you to create your own.

CHAPTER 10
Diminishing Involvement

Problems that need solutions are ever on the increase all around the world. New solutions also create new problems which will, in turn, need attention. There is always a genuine desire to do more, help more, reach more or affect more. We often get confronted with reality because we discover our limitation that we can only do so much ourselves within the twenty-four hours that every individual has.

If we need or desire to do more than our maximum capacity or more within the twenty-four-hour day can practically allow, we need to engage help beyond ourselves or our two hands. It is necessary if the output of the work must continue after our last action. It is vital if we want to get the same result even when we are no longer there.

The project I explained earlier in this book is a work I did for about four weeks. I completed it and delivered the outcome, but much more than that, it was replicable in a short time. The involvement required is less than a minute to start the process, and it takes about 10 minutes to run. Anyone can do it, it does not have to be me, yet it produces the same result. I left the role soon afterwards to progress my career, but it was exciting to note that it still worked without me, and it produced the same result.

There were some other problems I solved using the same principle. We were a small team and had so many limitations. It was crucial to have a solution that would work with limited involvement. One of the systems we relied on, the service desk system, did not provide value to our customers. There were complaints about its efficiency. It was a shared solution between two companies, but we did not get the best we desired. It was complex to use, challenging to track progress, and requests were not getting to the personnel that could help. It affected the productivity of the members of staff as they could not get the help they needed.

We got an inexpensive solution within the tight budget, and I invested some time to configure it to work with limited human intervention. It helped and effectively handled the workload and also covered for the insufficient hands. It did so effectively and continuously. The result was noticeable within a short time as all new requests get automatically assigned to the right person, and we had calls resolved in less than one minute of being logged on the system. It worked as programmed and did so continually.

I do not have to be there. It just works. It works when I am not there. I finish my work for the day and go home, but it still works. It works when I am on holiday. It still runs when I left the role. It is the concept of continuous results without continuous involvement.

The work gets done every time. The same result, but my hands are free to engage in something else more important. I can solve some other pressing problems or save the company some money by not paying me for unnecessary extra hours. I can also have a meaningful life to do something else to add value to the world. Amazing!

I was in another role to manage the virtual infrastructure supporting over five thousand employees. I was excited at the opportunity as it will allow me to put these skills to work. The scale was large, and it was also multi-site. I applied the same principles from previous roles but using compatible tools to reduce involvement, and it worked. I could deploy virtual machines by depressing the enter key after a command line using PowerShell and walk away. In a matter of time, they are fully ready and usable. One colleague of mine once logged an infrastructure request, and he later saw me away from my desk some minutes later. I told him that what he requested was ready. He was wondering: 'When did you do it. You were away from your desk?' I told him to check. He was surprised upon confirmation, but I smiled; It works. I do not have to be there. It will produce the same result all the time.

So many times, I could run such commands that affect and alter the configurations of thousands of systems or objects while I walk away doing and engaging in other productive work. I only return to completed work done without my continuous involvement. It is replicable anytime and every time under the same circumstances to achieve the same results.

If we can do more with less, we can solve more problems faster, and the world will require fewer experts to achieve the results achieved by experts. If we can implement the concept of reduced involvement in our solutions, it will improve the quality of life.

Everyone around the world can get help or support for any problem regardless of their expertise or discipline. You can communicate with anyone around the world no matter the language or culture. When problems are solved this way, we have more for less and more people could benefit. You will not need to learn German to communicate with a German. You will not need to wait for a professional to mark your examinations and wait for days to get the results.

You will not need to wait for a doctor to wake up to get a diagnosis. You will not need to wait for a lawyer to return from annual leave to get advice. There is a solution that makes things simple, such that anyone can use it. It is automation.

Some may see automation as a disadvantage, as it will take over the jobs of some people. It is an advantage as it is helping to do more with less. It is not about the loss of skill or expertise, but expertise and skills are not even sufficient compared to the global demand in most cases. Some countries in the world have less than twenty physicians per one million people.[1] In some others, the patients, in some cases, wait for over two weeks and sometimes up to four weeks to see a General Practitioner.[2]

The experts can provide the knowledge and skill to develop a system that proffers solution to matching symptoms. It will help many people around the world that do not even have access to a physician. The physician is not there, but they enjoy the results as if the physician were there. The few experts can manage complex situations while they are helping others through automated systems.

The systems will continue to work every time. It means that patients could also get medical advice while the medical doctor is asleep or some assurance in an emergency or a prescription when the surgery is closed. In a twenty-four-hour period, more patients can get help from wherever they are.

Imagine that such a system is available with global access, which means that thousands or millions of people can get the same result or help from where they are. Even if a whole town does not have a physician, they could enjoy the services provided by such global solutions.

Many people die of allergies without knowing. Many live with pains and complications they cannot even explain. All they need is medication or advice to keep them alive and functional. They cannot get help because they do not have access to experts. It will not matter if the expert is physically there; what matters is the answers.

Whatever skill you have could be of great benefit to more people continually without your continuous involvement using the advantage of automation. You may be frustrated where you are because your skill is not valued, appreciated or required, but somewhere around the world, thousands of miles away from where you are, that is just what they need. You can serve the world from where you are with what you have got without your continuous involvement.

Your knowledge, skill, experience, exposure, challenges, victories, education, and even failures could help many people. It has happened to you and affected you one way or the other. Many more people are having or going to have a similar experience, if not the same. Your input could be of great help, and more importantly, it could be of great help to people continually without your continuous involvement. Do it once and let it keep on working.

◆◆◆◆❖◆◆◆◆

We need to engage help
beyond ourselves or our two
hands.

◆◆◆◆❖◆◆◆◆

CHAPTER 11

Some Concerns

There are so many fears about automation. Yes, there are many disadvantages, just like many other solutions. You may even know more than I do. There will always be disadvantages, but that does not stop the progress and efficiency of automation. Every solution introduces the possibility of another problem, but it could still be worth it if the benefits outweigh it.

If we stop using automated machines, there would be a challenge to meet the global demand for several products in many industries. Human beings would also suffer harm as they work in harsh and hazardous conditions.

If you stop using cars, you would be prepared to have a long way to walk. If you refuse to use the aeroplane, you may spend all your years travelling. If you refuse to use online platforms for business, banking, communication, research, marketing, and training, you may not be able to maintain relevance or probably limited.

The world is changing, and if you rely on people for services, you may be at a disadvantage if you do not understand the technology. If you are in a business, you may not be able to compete with competitors. If you are looking for jobs, your skills may no longer be relevant.

You will do well to consider every option; then choose what is best for you to help achieve your goals. If your products are going to be relevant, the use of the current technologies may be helpful.

It is not everything that can be fully automated. Some may benefit from partial automation, while others may not be worth the risk or investment. Automation might not work for you in some situations, conditions or environments. It probably may not be the best or viable solution based on some vital considerations or factors.

It is also possible that some people would disagree with some solutions or decide to do without them. You do not have to agree to everything, but you can find what you are comfortable with and use it to achieve your goal. There are numerous possibilities, and pursuing all could be a distraction, but you could focus on relevant solutions that can help you become more efficient and effective.

The cost of implementation could be an issue of concern, but it depends on the solution. In some cases, the initial investment could be high with potential long term benefits. The cost should form part of the considerations to determine the viability of the project. It should also include the cost of maintenance and running cost where applicable. In some other cases, it could provide instant savings or reduced operating cost.

It could be beneficial and disadvantageous. It depends on what you do with it and what you intend to achieve. It is crucial to ensure that you do what is best for you in every situation, considering every relevant factor.

There will always be a risk; change will always have implications. We do not stop living because there is a risk; we consider the level and determine whether it is low enough to accept it or too high and avoid it. You will do well to understand what you are doing and its implications.

CHAPTER 12
Some Solutions

There are so many benefits to deploying automation. It helps to enhance effectiveness and enables you to do more in a limited time and reduced involvement. It is not just for technology experts or professionals but for anyone that desires to maximise time by diminishing involvement. Many have used it for simple or complex tasks and processes. It could be of benefit to anyone that would embrace it. It helps one to produce more with less.

Many people are struggling to perform operations that consume a lot of time and human energy. It has made many grow old so quickly. Hard labour is reducing their effectiveness and efficiency. They usually have little to show for it. They probably could do more, with more results, if they were able to implement automation.

We need food, but many have invested their life in agriculture, exposing themselves to various challenges. They survived, but some have been worn-out by the process. The risks and the challenges mean they may only have a little to show for their hard work. They often struggle with many things, including harvesting or processing their produce due to insufficient hands and resources. They could lose part or all of their harvest, with often significant financial implications.

It could explain why many in the younger generation do not see it as attractive. The number of people that work in agriculture has dropped to almost half in the last 30 years.[1] Introducing automated systems would not only attract the young and bright but also affect millions of people positively.

Many people interested in agriculture are frustrated by so many things. Their life could be much easier if they have relevant technologies to help. You could probably imagine the frustration of farmers as they helplessly watch their crops or animals perish. The yearly losses in the industry are staggering. In the US cattle industry, there is a loss of $1.5bn every year because about 2.5 million cattle die. They could avoid the loss if they could find a way to accurately and quickly monitor the cattle. A solution that will reduce their involvement, yet get the work done, is all they need. The development of automated solutions could turn the losses into profits.[2]

It would be fulfiling to have reduced involvement and still get the job done. There are many agricultural pieces of machinery set to start and complete operations. It goes a long way in making life easy, reducing the risks, cost and labour. Can you think of a scenario where one person runs a farm, and the person still has time to hang out with friends? It is possible using automation.[3] One machine effectively replacing several human efforts and completing the tasks in a short time. It also increases accuracy. We can do more by engaging automation, freeing ourselves to do the more crucial things.

You probably have heard of injuries and deaths in the constructing industry. It is usually due to the dangerous tasks human beings have to do daily. Machinery capable of replacing some of the involvement which poses a risk is available. There is room for more solutions, improvements and deployment. It helps achieve a faster result with improved safety in construction work like bridges.[4] Many companies are now able to do more with reduced danger to human life.

The time taken to complete processes reduces when an automated solution is in place. There is a piece of agricultural machinery. It has a working width of 33.5 foot and can operate on forty acres of land in one hour with no human involvement. An autonomous robot that does vegetable mechanical weeding, completing up to 12 acres in 9 hours.[5] I wonder how many people will be required to do the same and how long will it take them. It is equivalent to 9 football fields, or 180 tennis courts. The number of people that would complete it within the same time would be multitudinous. Can you imagine only a few hands manually doing such work daily? What will be the quality of their lives after 20 years of doing the same thing? Doing more with less will help improve their lives; they can do more with less.

It is possible to produce more and meet the needs of many more people by deploying relevant solutions. A piece of machinery can work unattended, at a constant speed, as much as required. There is leverage to produce more. It is interesting to see a piece of machinery that works unattended to complete a process that delivers package products ready for the stores. It is just amazing. It takes care of the whole process including, planting salad seeds, transplanting, harvesting, weighing and packaging.[6] We can do more, produce more with less, thus meeting the ever-increasing demand for food.

A dairy farm owner replaced manual equipment with automated systems. He could manage about two hundred and sixty cows all seasons of the year with the technology. The farmworkers required to feed and milk the cows, which are usually challenging to find, now have a viable replacement. The systems help with the mixing of the feed, making the feed available within easy reach. It does so as required in any of the twenty-four hours of the day.

The second part of the automated system can identify each cow and collect about a hundred and twenty pieces of data. The data has many uses, among which is to get information relevant for feeding and milking. As the cow enters, the teats get cleaned with a brush, a milking unit is attached, and the milking process commences. The farmer now has only four people on his payroll that are probably involved in other activities.[7] He found a solution for his unique need that will translate to more returns on his investment and less reliant on labour that is not readily available.

It is often challenging for businesses to be competitive. The cost and scale of production could affect their performance. Anything that helps you produce more with less may give that advantage required to compete on a global scale. Harvesting peanut and processing it with modern technology, getting different products in the process is a possibility that could represent a big win.[8] The processes complete with little involvement. It is a solution that can help mass production and distribution in many countries. There are wastages in many farms because of the low processing ability. The use of automated systems could help in the processing of raw material to finished products.

The use of automation technology can create opportunities. It can free humans from dangerous, repetitive, unpleasant, and laborious work. Many lives have been lost or hurt because of work in some very high risks conditions. Automation solutions could help prevent such loss and damage to human life.

It can also produce significant benefits and generate valuable opportunities for the present and future generations if applied wisely and effectively. It could provide growth and development that could enhance the quality of life and standard of living.

I hope you will use the principles to provide a solution that could make life better for you and others.

CHAPTER 13
A Product of Time

Automation is a product of time: It is not magic. Everything that works on its own without intervention does so because someone invested time to make it happen. The hard work involved in programming and scripting is what defines the actions. It determines what happens, the order in which it does, and the circumstances under which it does so.

The absence of human involvement and interaction means there are a set of instructions in the desired sequence, inbuilt or incorporated, that does the control. It can continue to perform the same way as expected and desired using those set rules. It is the hard work of somebody. It may be once, and if done well, it can now begin to work itself using the set rules to function without any further involvement.

Automation is not encouraging laziness, but rather it is hard work and time investment such that you do not have to keep spending the same length of time and effort only to get the same result you had once before. Moreover, it does not make sense because you are using your energy, time, and resources to do the same thing, sincerely expecting the same outcome, and achieving the same result.

You can save time which could help in doing more important things. If you can discover how to invest time such that the time invested into a product incorporates a solution that could make it work itself to achieve the same result, you have already achieved something with your hard work.

Think about it. If your involvement will not change, the process remains constant, and the decisions remain the same; it will continue to work. The incorporated program or rules, working with the relevant devices, can follow the sequence of operations. It can always perform the required actions to produce the desired outcome.

I know of a man that read a book of over 1000 pages, about 800,000 words in his local language. It will take about 70 hours and 40 minutes to read the book, so he did that once with his own time and preserved it in a media that can replay the same thing to achieve the same result. He did it once, but after that hard work, many people can continue to benefit from his work with their own time but without the involvement of the one that made it. He provided an educational solution for the benefit of those that understand the language. Everything that works by itself is a product of someone's time.

In programming, there is a system of measurement that measures the size of a computer program. It counts the number of lines in the text of its source code. It is known as lines of code. If you know its details for a program, you have known what it takes to make it work with no involvement. They are the commands or instructions that have replaced the interaction or involvement. Some people have inputted so many lines that they can comfortably sit back, relax and allow their work to run itself.

Let us consider some interesting statistics:

The average iPhone app has 40,000 lines of code.

Ubuntu has over 50 million lines of code.

Facebook runs on 62 million lines of code.

Photoshop C.S 6 has about 4.5 million lines of code.

Linux kernel 2.6.0 runs on 5.2 million lines of code.

Google Chrome has 6.7 million lines of code.

Firefox runs on 9.7 million lines of code.

Android has 12 million lines of code.

Linux 3.1 runs on 15 million lines of code.

Microsoft office 2001 has 25 million lines of code.

Windows 7 runs on 40 million lines of code.

Google has 2 billion lines of code.[1]

It takes a lot to run the software behind Google's Internet services—from Google Search to Gmail to Google Maps. It is a massive investment of time that keeps it running with reduced involvement.

If you power on your computer running Microsoft Windows operating system, you will discover that some actions happen before presented with the login screen. A lot of work and time made that happen. It is a set of instructions and operations, happening in sequence just by depressing the power button. It happens every time you do that, but Microsoft made it possible.

It could sometime be pages of commands and instructions that have to be incorporated into the product to make it possible to achieve the same result with just one action. It is usually hidden most times from the view of the user. You get notified politely to wait while the tasks go ahead.

Linux based operating systems also have a start-up sequence, which is a series of processes that takes place behind the scenes from the time you depress the power button until the Linux login prompt appears. You can see some progress and updates of the processes. It is those quickly moving texts on the screen, but with no action required. It is following a sequence of instructions, inbuilt by someone's precious time.

Have you ever wondered what happens behind the scenes when you touch your credit/debit card at the till to make a payment or when you input your card details on a website to complete a purchase? All you have to do is touch or type in the details, but the resulting actions, are planned to happen in a sequence to produce the same result repeatedly.

The hard work that generates that result all the time is a product of time. You got to the checkout, touched your card on a contactless card reader, and you got home to discover your money is gone. Someone invested time to make sure it happens that way even if he is not there. All those involved in the transaction will not even be present or visible to you.

I have written many scripts that are just groups of actions designed to run in a particular way. It follows the sequence even in my absence because I have invested my time to define how I want it to work, and with the relevant tools, it continues to work that way. I have used my time once to define and determine the process.

You can also invest your time into a solution that will continue to work even in your absence, producing the same result. You can convert your skills, expertise, experiences to solutions with your time such that many others without the skills or experiences can use their time to run it and benefit from it.

You use your time to make an automated diagnosis solution as a doctor. One that can help confirm an illness and probably prescribe medications. You would have converted your expertise with time by doing so. It will continue to run and work for the benefit of many, but not with your involvement. It now runs and works with the time of those that use it. You would be, by doing so, providing the benefit of your skill to many people, without your continuous involvement.

The level of automation is dependent on the time you are willing to invest in the product. It could be partial or complete. A fully automated system benefits from a total replacement of human involvement using relevant devices and instructions that enable them to function independently. If you can invest the time, you can also enjoy the benefits.

You have enjoyed so many things. Some people made it easy for you to use. You can also do the same by providing solutions that will continue to work, even when you are no longer there. It is a product of time.

◆◆◆◆❖◆◆◆◆

The level of automation is dependent on the time you are willing to invest in the product.

◆◆◆◆❖◆◆◆◆

CHAPTER 14
The Future or the past

There is one desire and expectation common to all of the 7.8 billion occupants of the earth surface. It is the desire for the future. At one time or the other in life, everyone would have thought of it, many are thinking, and many more will still think of it. What would tomorrow hold? Everyone desires to know. What would the future be in my field? What is the fate of this business in the years to come? Does this trade have any prospects? Will this lucrative venture today still command the same interest and demand in the future?

Some have desired a better future, others imagined, some expected it to happen, some others worked it out, while many waited for it. You must have realised that you are where you are today because of what you did yesterday or in the past. It also implies that what you will be in the future is a function of what you do today. There is also another possibility, and that is a situation where you do nothing. Doing nothing at all, yes, that also has a future because you are likely to be a victim of the change initiated by others that did something.

You cannot change your past; your past has brought you to where you are now. However, you can do something now, which is the present which could then change the course of your future if you so desire.

At the beginning of the 18th century, most countries of the world were similar economically. For instance, the differences in the countries in Europe and Asia were not too significant. Today we have classifications of countries like the first world countries, second world countries and third world countries that describe how rich and powerful nations are. How did this change happen? Europe and America gradually implemented solutions that caused them to experience significant changes, and the future of Europe and American began to differ from that of Asia.

In the 18th century, many were farmers and hunters, either in Europe, America or Asia, and the reward for corresponding labour would have been the same in those countries.

In 1760, something happened in England that changed the course of history. It was the change of tools and processes used. The results produced benefited a few people. The transition to machinery and manufacturing guaranteed more products for more people, using less effort and people. It is the industrial revolution that started in England and Europe. The United States took advantage of this change also.

The effect of the change made Britain the world's leading commercial nation by the middle of the 18th century. It also had a massive impact on the population as every aspect of life was affected. According to the records, the people and the average income experienced an unprecedented and continuous increase. It was the first time in the history of the western world that the general population experienced a consistent rise in their living standards.

It means that hoes, cutlass, axes, spades, among others, were the tools used for agriculture in the 17th century in Europe and America. It limited their production, and they had to put in so much effort and just had a little to show, but a change that was initiated and applied created a future that we see today. If all the countries using hoe in the 17th century embraced the same revolution, they would have experienced the same or similar improvements.

Any nation that still seeks a better future for its people can take advantage of the change. The future they seek is already the past of some other countries that took advantage of it a long time ago. The tomorrow of one country or people may be the yesterday of another. The future of one is the past of another.

Some of the countries in Asia, after many years, began to embrace and implement change. It has eventually redefined their future. Any country or people that can adapt and implement change will see a corresponding result. Some have done it and achieved an outcome that you desire and admire. It implies that if you do the same thing today, you will get the same corresponding result. The future depends on what you do now.

There are now fully automated machinery and processes that handle the complete lifecycle of agricultural produce. It operates at industrial scales with limited human intervention. The activities include loosening the soil, seeding, special watering, moving plants when they grow, and harvesting, among other activities. There are also machinery and processes that handle the harvested crops and transform them to finished processed products fully packaged, ready for delivery in stores for consumption. An entire nation could experience transformation by providing food and employment using the relevant technology.

I was once an apprentice in a printing press many years ago. I was learning about and using a technology invented in the 14th century while living in the 19th century. You may never have heard of it, but it is true. It is letterpress printing technology. Every letter or character, made of metal, was arranged to form words, sentences, and paragraphs. The press robs ink on the raised surface of the type, which is then pressed against a paper to transfer the ink. It then produces the impression of the typefaces on it.

I had to learn and master the location of each character in the letter case. It was full of metal typefaces, each character type separated in the case. To form words, I had to look for each letter from the relevant type case that corresponds to the desired size and load them accordingly to a composing stick. I had to form words, spaces and sentences one character at a time till completion.

I get to the end of the composing stick and then transfer the completed composition carefully each time to a tray until the work is completed and later, set them for printing in the printing machine. I also had to clean up and then redistribute all the typefaces to their appropriate boxes in the type case. It is the process I had to go through all the time to print any document.

Imagine you had to pick up each letter on this page you are reading right now from a pile of characters separated in a letter case and arrange each one in a composing stick. How long do you think it will take you? I wonder. Today you can depress a button on your keyboard, and the characters appear on your screen or probably speak to a device, and the words converted for you with the help of voice to text software.

The output and results will be different, and the time consumed will vary. The future of those using the printing press technology is already the past of some other people. The difference in productivity, effectiveness and efficiency of using the two tools which are generations apart could be up to one thousand times.

The time, energy, resources, and personnel that would produce a printed page using the older technology could create over a thousand complete books today. It is achievable by taking advantage of the new and more efficient technology.

Are you still looking forward to a future where cars will operate without a driver? It is already the past for some. If you hope for a car ignition that can start remotely, that is already past. If your future is the past, you need to move faster as you are still behind. It does not mean you should always follow the trend. In your area of speciality or assignment, you can make the most of the opportunities available to enhance your productivity and reach out to more for less. Not all the technologies will be relevant to you or your purpose, as many may be a distraction.

You need to focus on your purpose and get the relevant tools to fulfil your goals. You can become efficient and effective by taking advantage of all the available resources to help you do more for less. If you are a mechanic and still want to be relevant and competitive in years to come, you will need to embrace and implement change.

If you are into Information technology, you need to change. If you are a farmer, you cannot compete with hoes any more. The technology of the 17th century cannot put food on your table in the 21st century. There are some skills or technologies of the 20th century that are no longer competitive. You cannot be relevant except you embrace change.

Car manufacturers now produce cars with little human involvement. We have hybrid and electric automobiles with all kinds of technologies. It is already the past of the manufacturer but the future of the buyers.

Laptops now come with new features, capacity, finger recognition, and voice recognition. The manufacturer thought of it, designed it, tested it and then made it commercially. It is already past for them. You buy it new, but it is your future.

If you decide to do nothing, you are heading for a future that others have worked out that will become a change that may make you a victim. It may affect you and differentiate you from the rest. You may still be doing more for less while others are doing less for more.

Automation is already the past for some, although it is still the future for so many. The amazing truth is that the change will affect you even if you do nothing. Many became jobless as professional typists because of innovation. They had years of experience in the traditional typewriters but never prepared for the future of computers. The change others made affected them because they did not change. They became the victim.

Many phone operators have no jobs anymore because of automated answering machines. Checkout staff members in stores had to look for something else because their job is no longer required. Self-checkout and payment systems replaced them. Many publishers and professional printers have lost their jobs to automated and electronic publication and publishing. Marketing professionals have lost their jobs to online marketing software. Assessors have lost their jobs to automated assessing and marking tools. Training organisations now have online platforms. The world is changing, whether you like it or not.

If you lose your job because of advancement and innovation, it is proof that you are worth much more, and you can add value because there is something you can do that no technology can replace. In other words, there are things you need to stop doing, and there are things you need to start doing. If you are not sensitive enough to discern, your employer or your employer's decision may be the help you need to unravel the truth. It is just a polite way of saying there are things you ought to do now, so you need to stop this job you are doing and focus on more important things. You may need to close one chapter to open another.

Remember, the actions of today determine the future. The future of one could be the past of another. The yesterday of one could be tomorrow of another. You have the power to choose now before being confronted with a future without choice.

The technology is there to help your effectiveness and efficiency. It will help in the pursuit and fulfilment of your goal and assignment. Research has shown that most people use less than ten per cent of their mental capacity or potential. It implies that if you have more free time, energy and resources, you would be able to do more and serve or affect more people. That is why you should spend more time on what only you can do and get free time by taking advantage of available tools and processes.

If the western world were still using hoes as in the 17th century, they would all have kept their jobs, and the nation and its population would have become poorer, inefficient and worn out. A change could mean a necessary temporary loss that will work out a future gain. Do you know that the ancient printing equipment I was using was expensive in those days and had great worth? Yes, but if you want to be relevant in the business, you have to let it go, it will not hold too much value anymore, and it cannot keep you competitive anymore. You may have to decide to lose so that you can gain.

Many families would have also worried and feared so much during the industrial revolution. Their well-known source of livelihood disappeared and their once helpful tools, ignored and thrown into the trash. They must have felt a sense of loss. If they refused to change, their situation would have remained the same.

The effect of the initial losses of jobs and tools is what the world is enjoying today. It has delivered increased productivity, efficiency, effectiveness, better standard of living, multiplication and many other benefits. The revolution created more time and resources and helped them produce more and affect more people. It was the future they prepared for yesterday. It is still the future many countries desire, but it has now become their past. The expectation of the western world is a function of the change they are initiating today. You can be part of that change so that their yesterday does not become your future, but their future is also your future, or better still, their future could be your past. It is wise to make the most of the opportunity so that

your tomorrow is not anyone's yesterday,
but your yesterday their tomorrow.

◆◆◆◆❖◆◆◆◆

The future of one could be the past of another.

◆◆◆◆❖◆◆◆◆

CHAPTER 15

The Right Choice

I hope you have learnt something new and probably found some information that could help you begin or progress your journey in the world of automation.

Automation has benefited so many people, and many more are engaging the same to improve effectiveness and efficiency. It can make a difference when understood and appropriately applied.

You have the power to choose how it works for you and how you decide to apply it. You may use it for simple tasks to help you achieve orderliness and peace of mind or complex tasks that can help increase your productivity and reach more people with less while you still have time for crucial things in your life.

The benefit you get depends on what you would like to achieve or the problem you would like to solve. It is crucial to look for a solution that would work for your unique situation and meet your needs.

Proper consideration is necessary in each case, and professional advice is crucial. The idea of automation is to solve problems. The solution should, therefore, be worth the investment, except the purpose, is different.

You can make things work for you while you spend your time focusing on other crucial things.

Bill Gates was able to produce products that could run with less involvement. The organisation—Microsoft—can continue without his continuous presence. He could afford to step down from his role as CEO[1] to focus on something different.[2] The organisation is still achieving the same or even better results.

Jeff Bezos of Amazon also stepped down as CEO to recover his time and energy to focus on other things. He did not have to be there for the results to continue.[3] It is the same concept of producing continuous results with diminished involvement.

Many others have applied the principles, recovering their valuable time to pursue what is crucial without affecting the results.

If you have been doing the same thing to get results, you will probably still need to continue doing the same to get the same result. You may never get the time to do anything else. All you get is the same result.

If you always need to get to the river to get water, there will be a problem when you cannot go to the river. You have been doing the same thing for a while; it is time to make things work for you so that when you cannot work, or choose not to, because of more crucial things, you can still have water flowing.

You can decide to change for the better by doing something to effect change. It is not what you know that matters; it is what you do that makes the difference.

Choose to make a difference.

References

Preface

[1] A. Calaprice, Ed., *The Ultimate Quotable Einstein*. Princeton, New Jersey: Princeton University Press, 2010, p. 474.

[2] "Google Search Statistics - Internet Live Stats," *Internetlivestats.com*, 2009. https://www.internetlivestats.com/google-search-statistics/ (accessed Oct. 19, 2020).

[3] Wikipedia Contributors, "Google," *Wikipedia*, Dec. 18, 2018. https://en.wikipedia.org/wiki/Google (accessed Oct. 19, 2020).

[4] Wikipedia Contributors, "YouTube," *Wikipedia*, Feb. 18, 2019. https://en.wikipedia.org/wiki/YouTube

(accessed Oct. 19, 2020).

Introduction

[1] R. Hart, M. Casserly, R. Uzzell, M. Palacios, et al., "Student Testing in America's Great City Schools:An Inventory and Preliminary Analysis," Council of the Great City Schools, Washington D.C, Oct. 2015.

[2] M. Mohsin, "10 Youtube Stats Every Marketer Should Know in 2019 [Infographic]," *Oberlo*, Jun. 27, 2019. https://www.oberlo.com/blog/youtube-statistics (accessed Oct. 20, 2020).

Chapter 1: Simple or Complex?

[1] Dictionary.com, "Definition of automation | Dictionary.com,"

www.dictionary.com.
https://www.dictionary.com/browse/auto
mation?s=t (accessed Oct. 26, 2020).

[2] Techopedia, "What is Automation? -
Definition from Techopedia,"
Techopedia.com, 2019.
https://www.techopedia.com/definition/32
099/automation (accessed Oct. 26, 2020).

[3] Cambridge Dictionary, "AUTOMATION |
meaning in the Cambridge English
Dictionary," *Cambridge.org*, Nov. 20, 2019.
https://dictionary.cambridge.org/dictionary
/english/automation (accessed Oct. 26,
2020).

[4] Wikipedia, "Automation," *Wikipedia*,
Apr. 16, 2019.
https://en.wikipedia.org/wiki/Automation
(accessed Oct. 26, 2020).

Chapter 2: The Search

[1] "The History of Tea Kettle," *In The Kitchen.* https://inthekitchen.org/history-tea-kettle/ (accessed Oct. 30, 2020).

[2] J. Elworthy, "Inventor extraordinaire - who stopped kettles boiling dry - gives £2.5m to Cambridge University for innovation professorship," *Cambs Times.* https://www.cambstimes.co.uk/news/inventor-extraordinaire-who-stopped-kettles-boiling-dry-gives-2-5m-to-cambridge-university-for-innovation-professorship-1-5195951 (accessed Oct. 30, 2020).

Chapter 3: Behind the Scenes

[1] CBS News, "Almanac: The first automatic door," *www.cbsnews.com*, Jun. 19, 2016. https://www.cbsnews.com/news/almanac-the-first-automatic-door/ (accessed Oct. 20,

2020).

Chapter 4: Learn from History

[1] Jean-Paul Rodrigue, C. Comtois, and B. Slack, *The geography of transport systems*. London▯; New York Routledge, 2020.

[2] I. Hutchings and S. Shipway, *TRIBOLOGY▯: friction and wear of engineering materials.*, Second Edition. Butterworth-Heinemann, 2017.

[3] G. T. T. correspondent, "Self-driving cars could be allowed on UK motorways next year," *The Guardian*, Aug. 18, 2020. (accessed Nov. 24, 2020).

[4] C. Drescher, "The 10 Fastest Trains in the World," *Condé Nast Traveler*, Mar. 27, 2018. https://www.cntraveler.com/stories/2016-

05-18/the-10-fastest-trains-in-the-world
(accessed Nov. 24, 2020).

[5] "This is the world's oldest known boat:
The Pesse Canoe," *PortandTerminal.com*,
Nov. 23, 2019.
https://www.portandterminal.com/this-is-
the-worlds-oldest-known-boat-the-pesse-
canoe/ (accessed Nov. 24, 2020).

[6] C. Larwood, "A tour of the world's
largest cruise ship," *BBC News*, Apr. 24,
2018. (accessed Nov. 24, 2020).

[7] Virgin, "Science of Hot Air Ballooning,"
Virgin Balloon Flights, 2014.
https://www.virginballoonflights.co.uk/scie
nce-of-hot-air-ballooning/ (accessed Nov.
24, 2020).

[8] R. Miquel, "Giant flying machines: 10 of
the world's largest aircraft," *CNN*, Apr. 09,
2019.

https://edition.cnn.com/travel/article/worl
ds-largest-airplanes/index.html (accessed
Nov. 24, 2020).

[9] The Telegraph, "Trickiest airports, best
views and aborted landings: airline pilots
reveal all," *The Telegraph*, Aug. 04, 2016.
https://www.telegraph.co.uk/travel/comm
ent/airline-pilots-answer-questions-about-
flying/ (accessed Nov. 24, 2020).

[10] O. Smith, "The confessions of an airline
pilot," *The Telegraph*, Jun. 24, 2017.
https://www.telegraph.co.uk/travel/travel-
truths/confessions-of-an-airline-pilot/
(accessed Nov. 24, 2020).

Chapter 7: The Value

[1] M. Novotny, "At $200 a cup, This Might
Be the Most Expensive Cup of Tea in the
World," Aug. 22, 2019.

https://www.esquireme.com/content/3816
6-is-this-the-most-expensive-cup-of-tea-in-
the-world/ (accessed Nov. 24, 2020).

Chapter 10: Diminishing Involvement

[1] WorldAtlas, "25 Countries With Limited
Access To Health Care," *WorldAtlas*, Jan. 25,
2016.
https://www.worldatlas.com/articles/the-
countries-with-the-fewest-doctors-in-the-
world.html (accessed Jan. 09, 2021).

[2] Haroon Siddique, "NHS patients waiting
over two weeks to see a GP, shows survey,"
the Guardian, Aug. 11, 2019.
https://www.theguardian.com/society/201
9/aug/12/nhs-patients-waiting-over-two-
weeks-to-see-a-gp-shows-survey (accessed
Jan. 09, 2021).

Chapter 12: Some Solutions

[1] The World Bank, "Employment in agriculture (% of total employment) (modeled ILO estimate) | Data," *Worldbank.org*, Jan. 29, 2021. https://data.worldbank.org/indicator/SL.AGR.EMPL.ZS (accessed Jan. 25, 2021).

[2] J. Hoagg, "NRI: INT: Autonomous Unmanned Aerial Robots for Livestock Health Monitoring - UNIVERSITY OF KENTUCKY," *portal.nifa.usda.gov*, Feb. 15, 2018. https://portal.nifa.usda.gov/web/crisprojectpages/1015403-nri-int-autonomous-unmanned-aerial-robots-for-livestock-health-monitoring.html (accessed Jan. 25, 2021).

[3] ZF Group, "Automated Operations for Agricultural Machinery: Cooperation with Lindner Tractors (EN) - YouTube,"

www.youtube.com, Oct. 26, 2017. https://www.youtube.com/watch?v=DnjU6 dcpvBE (accessed Jan. 25, 2021).

[4] Radioposad, "SLJ90032 Bridge Girder Erection Mega Machine - YouTube," *www.youtube.com*, Oct. 24, 2015. https://www.youtube.com/watch?v=zvuufB qp0_4 (accessed Jan. 25, 2021).

[5] NaLac Technique, "Cool and Powerful Agriculture Machines That Are On Another Level - YouTube," *www.youtube.com*, Jun. 14, 2019. https://www.youtube.com/watch?v=4igFHz -pcdM (accessed Jan. 25, 2021).

[6] Handsome Dan, "Farming Automation - YouTube," *www.youtube.com*, Jun. 09, 2017. https://www.youtube.com/watch?v=ys_Lgs m95MA (accessed Jan. 25, 2021).

[7] Bloomberg Quicktake, "How Robots Are Saving the Dairy Farm," *YouTube*. Mar. 09, 2015, Accessed: Jan. 25, 2021. [Online]. Available: https://www.youtube.com/watch?v=-XI4siKp-nU.

[8] Noal Farm, "How Peanut Butter Is Made, Peanut Harvesting And Processing With Modern Technology," *www.youtube.com*, Oct. 22, 2020. https://www.youtube.com/watch?v=_GME _EG_o78 (accessed Jan. 25, 2021).

Chapter 13: A Product of Time

[1] D. McCandless, "Million Lines of Code — Information is Beautiful," *Information is Beautiful*, Oct. 28, 2013. https://informationisbeautiful.net/visualizat ions/million-lines-of-code/ (accessed Mar.

17, 2021).

Chapter 15: The Right Choice

[1] BBC, "Bill Gates steps down from Microsoft board to focus on philanthropy," *BBC*, Mar. 13, 2020.

[2] B. Gates, *How to Avoid a Climate Disaster : the Solutions We Have and the Breakthroughs We Need.* Random House Inc, 2021.

[3] BBC, "Jeff Bezos to step down as Amazon chief executive," BBC News, Feb. 03, 2021.

Thank you

Thank you for investing in this book. I hope you enjoyed reading it as much as it has been a pleasure for me to write it, and I trust that it has helped you.

It would be helpful to me and many others if you could share your experience by way of a review or a comment.

Please write to:

iph@TheServantandKing.com

We would be glad to hear from you.

Looking for more?

Please visit:

www.TheServantandKing.com

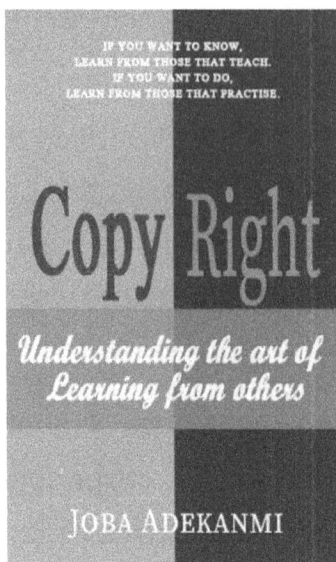

IF YOU WANT TO KNOW,
LEARN FROM THOSE THAT TEACH.
IF YOU WANT TO DO,
LEARN FROM THOSE THAT PRACTISE.

Copy Right

Understanding the art of Learning from others

JOBA ADEKANMI

Copy Right:
Understanding the Art of Learning from Others.

IF YOU WANT TO KNOW,
LEARN FROM THOSE THAT TEACH.
IF YOU WANT TO DO,
LEARN FROM THOSE THAT PRACTISE.

This book, with permeating shrewdness and practical stories, exposes the influence of other people. It will help you understand the impact on your life.

Available on Amazon or your local bookstore.

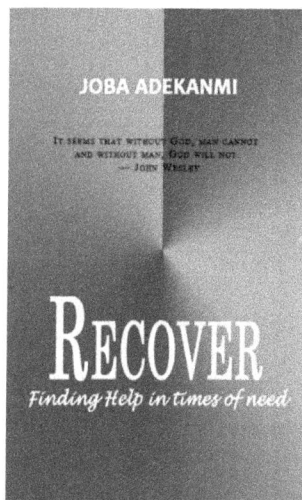

Recover: Finding Help in times of need.

IT SEEMS THAT WITHOUT GOD, MAN CANNOT
AND WITHOUT MAN, GOD WILL NOT.
— JOHN WESLEY

Recover, packed with real stories and recommended actions, will help you in your journey to navigate your present or future challenges to fulfil your potential.

Available on Amazon or your local bookstore.

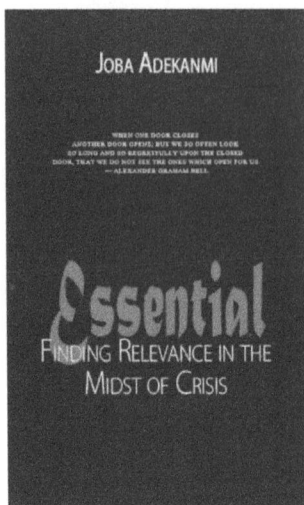

Essential: Finding Relevance in the Midst of Crisis

WHEN ONE DOOR CLOSES
ANOTHER DOOR OPENS; BUT WE SO OFTEN LOOK
SO LONG AND SO REGRETFULLY UPON THE CLOSED
DOOR, THAT WE DO NOT SEE THE ONES WHICH OPEN FOR US.
— ALEXANDER GRAHAM BELL

In Essential, you will find some clear and actionable advice that will help to restore hope and encourage you to step out and make the most of any situation. You are essential because you can make yourself relevant even when things change.

Available on Amazon or your local bookstore.

www.ingramcontent.com/pod-product-compliance
Lightning Source LLC
Chambersburg PA
CBHW072308210326
41519CB00057B/3090